SPINNING
OUT

SPINNING
OUT

*Climate Change, Mental Health
and Fighting for a Better Future*

CHARLIE
HERTZOG
YOUNG

FOOTNOTE

First published in 2023 by
Footnote Press

www.footnotepress.com

Footnote Press Limited
4th Floor, Victoria House, Bloomsbury Square, London WC1B 4DA

Distributed by Bonnier Books UK, a division of Bonnier Books
Sveavägen 56, Stockholm, Sweden

First printing
1 3 5 7 9 10 8 6 4 2

A CIP catalogue record for this book is available from the British Library
and the Library of Congress

ISBN (trade paperback): 978 1 804 44031 5
ISBN (ebook): 978 1 804 44034 6

Printed and bound in Great Britain
by Clays Ltd, Elcograf S.p.A.

MIX
Paper from
responsible sources
FSC® C018072

For the mad, the weird and the wild –
all those who dream awake

Table of Contents

'I am no longer accepting the things I cannot change. I am changing the things I cannot accept.'

Angela Yvonne Davis

'The civilised can pretend to be emotionally healthy as they do not commit genocide and destroy landbases, but instead take what they need to develop their "advanced state of human society". This is how we can all pretend to be sane as no one kills the planet.'

Derrick Jensen, *Endgame Volume 1*,
Seven Stories Press, New York, 2006

INTRODUCTION

Is it hot in here or is it just me?

I AM PRETTY MAD, by conventional standards, but I'm nowhere near as mad about it as I used to be. I first noticed the divergence nearly twenty years ago, when I was twelve. That was the year that the immensity of climate change properly entered into my awareness. It hit me like a planetary death drive, a bodily horror that made me want to disappear. I wanted no part in the pallid, hollow ways we lived, the suffering it entailed. Paradoxically, this new meaninglessness of the dominant culture somehow gave my life a charge, a frenetic, revolutionary meaning I had never felt before. It was an odd sense of belonging, especially in the lonely moments when I saw humankind as an ecological mistake. I was divorced from consensus reality and wedded to visions of utopian futures. To me, these projections became just as real as the material world everyone else thought I lived in. They were just as real, too, as the flickering visions of climate apocalypse.

That same year, as I was reeling from the realisation that the world was probably about to end, my family lived out its own private doom. My grandmother went missing. We put posters up around the Isle of Wight, where she had lived. Even on the pinned-up printouts she looked forlorn. My dad went from pillar to pillar, an abandoned boy. He told us she would be fine, trying to suppress the fear that his words were just consoling lies. It took six months for her to be found. She had jumped off a cliff.

Her body had been in the sun, the wind and the rain, and was found in a remote cove by a child exploring the coastline on a fishing trip. She was identified by her teeth and a metal pin in her elbow. I remember telling some friends at school that she had died by suicide. They told me she was in hell for killing herself.

As often happens for kids, these two dramatic episodes fused themselves in my psyche. Climate change has loomed over me ever since as an unpredictable harbinger of death. It's an embodied fear and fury that has led to a lot of mental unravelling. Climate change was a mythological counterpoint to a world I felt was pointless. At times it was a terrifying but exciting herald of apocalypse. Sometimes that gave me a powerful need to be an activist, to rebel. Other times it was an inescapable reason to destroy myself, just as we are destroying the planet – a rebellion of a different, darker kind. What was happening, in my formative youth, was the cresting of a disorder of the mind, fuelled by having to live on a fundamentally disordered planet; one that was threatening to shake us off its back for what we had done and for what we continued to do.

For years climate change was my everything. I started out giving talks in schools, pamphleteering, getting people to attend screenings of climate-change movies. I got involved in community organising and set up a green council at my secondary school. I somehow won a national award. At the ceremony I helped orchestrate a stunt from the toilets of 10 Downing Street, where a friend of mine superglued himself to Gordon Brown. Latching on to the Prime Minister, he shouted: 'You might be able to unstick my hand, but you won't be able to unstick yourself from the problem of climate change.' It was an unwieldy phrase, but it got the message across. The picture I took made the front page of the *Metro*. It was around this time that I started hallucinating properly.

One evening I was sitting on a mound on Hampstead Heath. The sun had just dipped below the horizon and the darkness encroached on the twisting Scots pines. I closed my eyes and imagined myself floating into the sky, skirting treetops and skimming along meadows of tall grass. There was something in the corner of my vision. A lumbering but agile form. It was quick and heavy in the blackness. It slowed and landed waist high in the grass. There was something flattening the foliage as it progressed from a few metres away, something cutting its way through. It stopped. I listened. The silence was dense. Then, like an explosion, the wolf pounced. I saw a flash of matted black fur and yellow teeth. I opened my eyes and I was back on the mound but a putrid stench hung in the air.

From that day on, often when I found myself alone and usually in the dark, the wolf would visit me. What started out as fleeting appearances, shimmers in the corner of my sight, became full-blown altercations. I would be walking home from school in the cold and turn a corner to find it lying in wait on the pavement, or curled up on the bathroom floor in the middle of the night. Its teeth would rip into my flesh as I kicked and flailed on my back. I would jab at its eyes, its gut, wrench its jaws open, anything to stop the attack, until it decided to lope away, like it was suddenly bored.

The wolf also liked to come on direct actions with me. When I snuck alone to the perimeter fence of a major UK airport, it trotted along beside me like a scout, then kept me company in a tree whilst I radioed dozens of protestors who broke in and locked themselves on, shutting the runway down. The wolf was with me, too, in the cavernous airplane hangars of UN climate meetings, in the hallowed wankery of the World Economic Forum in Davos, and lay on the concrete paving of Zuccotti Park during the dawning of the Occupy movement.

I had my first serious breakdown in 2009 after the collapse of the UN climate talks in Copenhagen. We had been told that it was the 'last chance to save the world'. The talks ended in disaster. I truly believed we had passed the point of no return; I collapsed and didn't leave my room for a long time. The wolf stalked the perimeter of my bed with rank breath. The pyromaniac in me, that sliver in all of us that revels in destruction, stirred. It felt wrong, but in that private moment, alone after the declaration of apocalypse, I wanted to see it all burn.

We didn't deserve the planet – we had been given a chance to save ourselves and squandered it, rich nations unwilling to compromise on economics to save civilisation. Those in power in the Global North were clearly too distant from the natural world, prioritising imaginary games over survival. At that time I managed to pass for sane, studying and working proficiently, but I lost some of my anchors. I was more scared, more delicate and more obsessed with working furiously to stitch my visions into reality. At the age of nineteen, after five sleepless nights riddled with hallucinatory imagery, an incandescent breakdown led me to a psychiatric unit. I was diagnosed with bipolar disorder.

In the intervening decade I have been in and out of study, in and out of work and in and out of hospital. It's been tempestuous. The wolf disappeared as soon as I started taking lithium. I swapped manic incandescence for deep depression. I still see things. Long ago it was a gangly spectre that lived in the corner of the room I barely left for two years. I also thought I was being followed for a long time, especially when I lived in a warehouse on an industrial estate. That led me to ditch my phone and draw maps for myself in code so I couldn't be traced. I then, obviously, kept getting lost. I had to sleep in places I shouldn't have. Sometimes I saw vivid sparks and trails of light when cycling through the city in the middle of the night. Sometimes I heard people talking about me in the walls. I became an addict. I lost my memory, my drive, myself. I had electroconvulsive therapy dozens of times, inducing seizures to kickstart my broken brain.

I eventually ended up jumping off a six-storey building. In stereotypically melodramatic fashion, although I have no memory of it, I fell from the flat roof extending from my then-therapist's office. I landed on the concrete base of a neighbouring garden. I somehow remained conscious and even had a long chat with the woman who found me. She thought I was a burglar, propping myself up on the planters and talking rather pleasantly – I'm told – as I bled out. I would have died, had an ambulance team not happened to be on the next block. My pelvis was split open and shattered, my legs demolished and my wrists badly snapped, and my legs so damaged that they both had to be amputated. I spent a month in a coma and six months in hospital. I was spat out into the pandemic, with lockdown hitting a month after I moved back into my mum's house to recover. I had lost a job, a flat and my then-partner. It took time to start actually recovering myself, and even longer for me to open up enough to realise how deeply traumatised my family and loved ones were by what had happened. I am still learning to balance my own needs with how I can be of most service to them, a process I imagine will take longer than a lifetime.

Since my diagnosis I have had to become a different person, especially since I jumped. The experiences of madness and the multifaceted recovery I have worked through, and continue to work through, have taught me a great deal about what it is I care about and how it is I want to be. They have also shed a piercing light on the disordered state of the world. Mine is just one case that demonstrates some of the interactions between climate catastrophe, society and the mind.

How we fit ourselves into this story is key to fighting climate chaos and the mental-health epidemic, let alone winning. This book is about figuring out how we are embedded in the dominant culture that is responsible for these crises, how we fit, and how to free ourselves from it. Social transformation is obviously and undeniably challenging, but if we understand its tiered nature – its benefits for us personally as well as systemically – then we can form a different relationship to this kind of work. Engaging in the right way also makes it less draining, and therefore more sustainable.

Recently, a close friend of my mum's, a counsellor, asked me why it was I had wanted to kill myself. We talked a lot. Eventually she turned to me and said: 'It sounds like you wanted to turn all your anger about the world in upon yourself, and ultimately towards the people who love you.' I nodded. It was hard to hear but I couldn't deny it was true, conscious or not. 'But,' she continued, 'what would happen if you turned it towards what you're really angry at instead? Towards the system that has caused all of this. Wouldn't that be powerful?'

We need to fight back, urgently. But we also need to get ourselves into a headspace where we can be well, stay well and use what is being done to us as a mobilising force. Some clinicians use the term post-traumatic growth, which I like. This is the idea that some people, after trauma, can enter a state of systematically rebuilding their lives anew. What I particularly love about it is that it focuses on recovery through the realignment of one's relationship to the world. In so doing, we recover not by detaching from society and just looking after ourselves, but by reweaving our relationships and using them to create the world we want to see. We need post-traumatic growth on a civilisational scale.

The Uruguayan poet-historian Eduardo Galeano once wrote: 'Utopia is on the horizon. I move two steps closer; it moves two steps further away. I walk another ten steps and the horizon runs ten steps further away. As much as I walk, I'll never reach it. So what's the point of utopia? The point is this: to keep walking.' For someone who thought he would never walk again, every step I've taken since is an embodiment of active hope. On my good days, setting my prosthetic on the pavement can feel like the beginning of a revolutionary act. I don't want to walk that path alone.

PART ONE

1

What we talk about when we talk about eco-anxiety

'We have an obligation to be anxious, it's a mark of respect for the gravity of the situation.'

Mark Corrigan stuck in the 'Nether Zone':
S7 E4 of *Peep Show*.

IT IS EASIER, AS the saying goes, to imagine the end of the world than the end of capitalism.[1] We are now being forced to stare down the barrel of the end of the world, whilst being told that everything needs to stay essentially the same. Our dominant culture is one of extractive, individualistic capitalism, built on age-old traditions of disconnection and domination. It's a system that is horribly adaptable and resilient, managing to maintain the outdated social structures responsible for the polycrises we are enmeshed in, whilst at the same time paying lip service to global catastrophe. This is a form of psychological sadism. It only succeeds by warping collective reality, depriving us of communities of care and cleaving away any possibility of systemic alternatives.

To live within such a society requires us to maintain what psychologists call cognitive dissonance. To function, whilst identifying as relatively powerless, we must simultaneously hold two incompatible realities in our minds or else deny one of them. Denial requires us either to ignore the physical realities of the climate, or else to detach ourselves from the dominant culture and try to manage private visions of existential doom.

In any discussion of climate change and mental health it doesn't take long for eco-anxiety to come up. It is what everyone mentions first. Eco-anxiety is a trendy, limited, tangled mess of ideas, full of false dichotomies and internal contradictions, but it is a useful term nonetheless. It's also, somehow, cool. *Grist* magazine recently dubbed eco-anxiety the year's top 'pop-culture trend'.[2] It was ushered into Anglosphere officialdom in the run-up to the 2021 UN climate talks in Glasgow, when the *Oxford English Dictionary* added the term to its lexicon. Over the course of that year, online searches for 'climate anxiety', virtually interchangeable with eco-anxiety, ballooned by 565 per cent.[3]

Eco-anxiety is the tip of a rapidly melting iceberg. It is also a crucial entry point helping a lot of people to understand the messy web of relationships between ecology and the mind. Eco-anxiety can be worry, but it can also trigger or be defined as a serious psychological disorder. Global authorities describe eco-anxiety variously as an 'unease or apprehension', a 'concern' about environmental destruction and a 'chronic fear of environmental doom'. Some think it should be categorised as a psychiatric disorder. Others say no, that it is a healthy psychological response to systemic collapse.

Generally, at least in the medical research community, eco-anxiety is seen as degrees of distress, rather than an explicit 'condition'. It shouldn't be pathologised, they say, and shouldn't usually necessitate psychiatric intervention. If we pathologise an appropriate emotional response to ecological devastation, then there's a risk of individualising a problem that is really systemic in nature. So, are these categories even useful? Do they change our experience of the world? Do they provide solace and support?

Often, the answer is yes. Many of the people I interviewed for this book have found the idea of eco-anxiety incredibly useful. It gave a name to something they knew, but had no words for. It gave them something to hold on to in a sea of uncertainty. Jennifer Uchendu, founder of The Eco-anxiety in Africa Project, says that her work on climate change was relentlessly unsettling and confusing until she found out about eco-anxiety and had a framework to help explain her feelings and those of her community. Eco-anxiety is a stressor, even if it isn't a 'disorder'. But it's a pervasive stressor which can easily push people into serious mental-health issues.

Eco-anxiety is usually associated with young people in the Global North. That is largely because the research funding is there, as is the global media focus, and mental health is often perceived as a Western luxury. But people in the Global South are already more acutely affected by climate-related mental-health issues. In the Philippines, almost 50 per cent of young people are 'extremely worried' about climate change and 74 per cent say that thinking about it makes day-to-day functioning difficult. In Brazil, half are struggling to function because of eco-anxiety. In Nigeria, it's two-thirds. In India, three-quarters. For the UK, the US and Australia, by comparison, between a quarter and a third of young people report similar difficulties.[4] That is still a lot, and many are not sleeping, eating, or working. People are struggling to find joy, connection and meaning in their lives because of their disintegrating future.

One UK study used a tool called the Climate Anxiety Scale, designed to pinpoint and define more alarming levels of eco-anxiety. They found that around 4 per cent of people in the country, of all ages, were suffering from moderate to severe symptoms. Respondents were having nightmares, dissociating and having difficulty connecting with loved ones. The authors described some of the cases as 'disabling'. Between 2020 and 2022, they found an increase of 15 per cent.[5] A similar paper, this time from the US, found that up to a quarter of Americans could be similarly afflicted.[6]

There is a common feeling of panic when we start to fully comprehend the immensity of the climate crisis. Everything is going to change. Ro Randall, a psychotherapist who specialises in climate change, talks about there being two stages in the experience of climate distress. First, there is the shock of realising its scale. This stage is extreme and incandescent. We can end up having to deal with fear, dread, feelings of unreality, anxiety, depression, psychosis and many other difficulties that fall anywhere on the spectrum from negative emotions to serious mental-health issues. The second stage is a slower burn, the long tail of coming to terms with this new reality we find ourselves in. This can be even harder to deal with than the initial realisation. We may end up questioning who we are, fundamentally reanalysing what we want out of life, and the nature of the system we live in. This can be destabilising, and it is incredibly difficult to deal with alone.

Feeling powerless and not trusting those in power to do anything meaningful about this hulking, existential, civilisation-level danger results in a perfect storm of mental distress. Around 40 per cent of children aged between eight and sixteen in the UK, for instance, don't trust adults to deal with the climate crisis. Some 20 per cent of children surveyed say that climate change is giving them nightmares.[7] Ecological destruction is crawling out of kids' subconscious when they are most vulnerable: asleep, alone and in the dark. Young people are used to living in a state of persistent disquiet, so we are very susceptible to eco-anxiety. We, as young people, are going to see the worst of climate chaos. On top of that, we grew up with comparatively little economic security in an era of persistent global crises. This was complemented by the distractions and demands of hyperactive technology and, in much of the world, deteriorating social safety nets. We are already more likely to experience mental-health issues than the rest of the population. In the US, even before the pandemic, one in three young people suffered from persistent feelings of sadness or hopelessness, whilst one in six reported having made a suicide plan. In both cases these represent 40-per-cent increases on the previous decade.[8] In the UK, the likelihood of a young person having a mental-health problem increased by 50 per cent over the three years to 2022.[9] In China, the number of young people suffering from depression increased between 2016 and 2020 by 35 per cent – and has reportedly shot up again since the pandemic.[10]

Understanding our internal worlds

Amongst young people, eco-anxiety is perhaps the main way to describe the effects of climate change on mental health, but, as a term, it is absurdly limited. There are myriad different descriptive terms that are often used interchangeably when it comes to the impact of climate change on mental health. In addition to eco-anxiety, there are also eco-anger, eco-paralysis, ecological guilt, climate anxiety, climate grief, climate depression, climate-change delusion and apocalyptic fatigue. How all of these relate to more conventional diagnoses of mental-health conditions is still a site for exploration.

As a starting-point, there is an urgent need for us to better understand our internal worlds. We know there is a worldwide epidemic of mental ill-health. Mental-health issues are already the cause of almost 15 per cent of deaths worldwide. Depression is now the leading cause of disability in the world. The burden of mental-health conditions has worsened in every country in which it has been measured over the past thirty years, and deteriorated further during the pandemic.[11] Climate change is already making this worse, according to both the UN and the World Health Organization (WHO).

I recently heard from Kerry, who told me she was struggling to cope because of climate change. She got in touch with me after I ran a poll asking people whether climate change has been detrimental to their mental health. Almost 2,000 people responded.[12] I found it astonishing that 79 per cent of people said climate change had been bad for their psychological well-being; 59 per cent reported a 'moderate' effect whilst 20 per cent said it had been 'severe'. One of these respondents was Kerry.

> I consider myself to have robust mental health, having been through quite a bit, including infertility, miscarriage and divorce. I tend to be very resilient and bounce back to the optimistic side before long.
>
> But ever since I found out the true magnitude of exactly how fucked we currently are *vis-à-vis* climate change, and

the appalling gulf in actions from governments everywhere, my mental health has taken a huge beating.

I have a toddler son and I spend at least part of every day thinking about his future and mine, how to escape, where might be safe. In the worst phases, these thoughts get very bad and everything seems pointless. I think about how I could possibly end things painlessly for my son and me if the situation gets desperate.

These are the thoughts I have every day. They are intrusive and unrelenting.

Another young woman I interviewed told me she felt 'deeply dissociated, anxious and depressed' because of the climate. She told me she had had to remove herself from activism because it made her feel like she was 'living in a simulation.' Now, she had to survive by living a 'smaller life' which nonetheless made her feel 'incredibly guilty,' both because of her own emissions and because she wasn't doing more to address the crisis. 'More than anything else it's the anger,' she told me, anger that she has been forced into this position by people more powerful than her. She will not be having children, along with most of the people her age with whom she has discussed it. For her, it is because:

> I don't trust that the world will be safe enough once I am gone . . . I don't think humans are kind in crisis. When one half of the world is being destroyed, I don't think the others are simply going to welcome them. I used to think it was worth doing something, when I thought some politicians at least seemed to care. Now, why would I do anything when they're not? I feel nihilistic.

Deciding whether or not to have children is a common offshoot of eco-anxiety and eco-concern. It reminds me of an episode in Mieko Kawakami's novel, *Breasts and Eggs*. Two women are sitting on a bench in a park discussing having a child. 'Why,' Yuriko asks, 'do people see no harm in having children? They do it with smiles on their faces, as if it's not an act of violence. You force this other being into the world, this other being that never asked to be born ... You think I sound extreme,' she continues, 'or detached from reality.'[13] When I first read this, I did think it was a bit extreme. Now, four in ten young people globally say they are not going to have children because of climate change. Nearly two-thirds of young people in Egypt say the same, along with more than half those surveyed in South Korea, India, Thailand, Hong Kong and Turkey.[14]

Many of the activists I started working with around fifteen years ago have either had breakdowns and left the movement, or have changed focus because perpetually being on the front line was too intense. Those who suffered most, especially those with subsequent mental-health struggles, were usually involved in direct actions, long-term occupations, and had been arrested or abused by the authorities. Back then we were not well practised in how best to support each other psychologically – though the situation today is not much better.

One activist I spoke to, whom I met in a temporary protest community, told me that they had not been able to think about anything but climate change ever since she realised how huge the issue was. She said that the kinds of activist and organiser spaces she moves in now can be overwhelming, largely because of an inundation of all the other issues tied to climate change – poverty, abuse, migration, racism. They get tied together, systemically, and they sprawl. It can feel like too much to handle.

She said that exiting these spaces, going into the outside world, where everyone else lives, then makes you feel atomised. It is easy to assume that everybody thinks you are mad, even if you know deep down that it is actually society that is sick. The experience is alienating, to say the least. A lot of people who live and spend time in activist communities have mental-health issues, often because of the way they are treated by society. These communities often tend to be very aware of wellbeing, but, as this activist shared with me, along with many others, there is neither the social infrastructure nor the skills, knowledge and tools to cope with the psychological ramifications of what comes up. Even in these aware and caring enclaves, bubbles pinched off from the mainstream, we don't know how to live, let alone protect our minds from dissolution.

Those who believe in the imminence of climate catastrophe can be forgiven for collapsing. Climate change can strip the world of meaning, especially when the newly perceived extreme reality is contrasted with a widespread generalised apathy. It can undermine every other facet of life. What is the point in studying, working, or building a family if the world is going to collapse? We have evolved to cope, by and large, with the knowledge of our own mortality. But Armageddon, death on a planetary scale, is something we have rarely had to contemplate within our lifetimes. It has happened before, during World War Two, the Cold War and, for some during the pandemic, all of which had serious mental-health consequences. On a smaller scale, end-of-days cults wrestle with annihilation, and many of them, too, suffer from more than their fair share of psychological difficulty. As human beings, we tend to resist apocalyptic narratives, even if they are backed up by fact. Studies show that most people underestimate existential risk, even when they are explicitly shown how dire things could become. According to psychological experiments, the only people found to assess unfavourable predictions accurately are those with depression.[15] Others have a comparatively delusional outlook.

Climate catastrophe is different from previous 'doomer' scenarios. First, there is a mismatch between our feelings of guilt and responsibility and our capacity to do anything about it. We feel responsible, and yet impotent. Climate change is also a spectral crisis, in comparison to, say, nuclear apocalypse. It is creeping and cumulative, and there is no clear binary of winning or losing the battle – despite the use of words like collapse, catastrophe, breakdown, apocalypse, emergency and chaos. Climate change is pervasive, too. We all live in the climate, as well as being responsible for emissions. We breathe the climate in and out of our lungs every three seconds or so. It is also nebulous, in that it's impossible to say for certain whether any specific activities are responsible for any particular climate impact. This malleability of cause and effect is a breeding ground for toxic imaginaries.

Our need for meaning

In all, what climate change does is eradicate the likelihood of a stable future. This is dismal, though not definite, but it nonetheless makes it difficult for people to have a sense of purpose, knowing that everything could fall apart. In 1946, the logotherapist Viktor Frankel wrote a book about his experiences in a concentration camp called *Man's Search for Meaning*. In it, he argued that meaning, for human beings, is nearly as important as food and sleep. Research suggests, explicitly, that a lack of meaning combined with the 'existential' threat of climate change is responsible for driving significant numbers of people into depressive states.[16] Other scholars argue, convincingly, that some forms of psychosis should be seen as evolutionary adaptations, coping mechanisms that explicitly search for (and create) meaning in uncertain and disturbing situations.[17] These states may have some function. They may be trying to tell us something important. Meanwhile, healthy resilience practices are being undermined. Even though spending time in nature is beneficial for our mental health, people who care more about nature appear more likely to be anxious and depressed, the cause being distinctly related to the destruction of the biosphere.

In *Living in the Borderland*, Jerome Bernstein outlines a discussion with a patient suffering from depression, isolation and despair. The patient had seen a cattle car and was deeply disturbed that the animals were on their way to slaughter. Bernstein suggested that it was a case of projection, that the cows represented something else to her, but she argued with him. 'It's the *cows*!' she said. Later, after encountering some stray dogs, she was similarly distressed. Again, Bernstein tried to find out what was *really* going on. This time, 'out of character for her, she became angry – so angry that she took her shoe off and hit the floor with it. "You just don't get it!" she shouted, and slammed the floor again with her shoe. "It's the dogs!"' Bernstein realised that his 'standard interpretations were not enough and somehow off the mark'.

Sometimes, the experience of contact with nature's demise is enough, on its own, to tip us into despair. There is no psychoanalytic analogue, no past trauma being forced to the surface. What is occurring, the thing we connect to directly, is the cause itself.

This may have been the case when a seventeen-year-old Australian boy was diagnosed with the world's first case of 'climate-change delusion'.[18] He was admitted to hospital with symptoms of depression, insomnia, guilt, worthlessness and suicidal thoughts. He was convinced that, due to climate change, his own water consumption could kill 'millions of people' through the exhaustion of scarce supplies. He stopped drinking altogether. He was unable to acknowledge that the belief was unreasonable, even when challenged. At the time, young people in Australia were significantly more concerned about climate change than they were about terrorism.

During one period of mania-induced psychosis I hallucinated strong winds which over days built to a hurricane. I believed I was responsible, that it was my duty to quell it. I was in Brittany, and my dad came from London to get me, worried for my safety. On the route home I found I could control the wind with my breath. When we got onto the ferry, the boat's rocking convinced me that people on it were going to die unless I went up on deck and blew the charged air from my lungs into the sky. I suddenly felt, knew in fact, that I needed to stand on the very tip of the ferry's bow. As I leaned and swung one leg over the guard rail, I saw the curved blue slope of metal in front of me fall away into a raging sea. My dad grabbed my wrist and shouted for help. Another passenger – my dad thinks she was an off-duty police officer, considering how she carried herself – managed to persuade me to climb back over the rail and return to my cabin. On the drive home I was convinced that I had failed, that I had killed people. I sobbed inconsolably, and found any attempts to comfort me deeply upsetting. It proved that they couldn't understand, and that I was all alone.

In *Mourning and Melancholia,* Freud wrote that depression is an expression of anger and aggressive impulses turned in on oneself. Some papers suggest that getting angry instead of sad can have a drastically positive impact on our ability to take action, rather than fall into despair. Others say eco-anxiety can be a strong push into activism.[19] It is not always a matter of choice, though – and presenting it as such can lead to self-blame if we don't have the capacity for active recovery right now. But turning negative climate-related emotions inwards, rather than aiming them at their rightful targets, can sometimes be fatal. In 2018 a retired LGBT rights lawyer named David Buckel set himself on fire to protest inaction on climate change. A lifelong conservationist, he had railed against the gutting of the US Environmental Protection Agency, announcements of new oil and gas licences and the deprioritisation of climate policy under the Trump administration. Before he self-immolated, Buckel sent an email to media outlets stating that his action was symbolic of the early deaths so many people had suffered due to climate change. Poetically, he used fossil fuels as the accelerant. He left a note for the first responders: 'I just killed myself by fire as a protest suicide . . . I apologize to you for the mess.'

Almost exactly four years later, on 22 April 2022, Earth Day, another American set himself on fire to protest climate inaction. Wynn Alan Bruce, a fifty-year-old Buddhist and climate activist, sat in front of the US Supreme Court and burned to death. It was difficult for me to decide whether to include these accounts in this book for two reasons. First, I don't want to glorify the actions of these men, much as I deeply respect their sacrifice. Nor do I intend to encourage people experiencing mental-health issues connected to climate change to put themselves in anything like this position. Instead, I raise the stories of David Buckel and Wynn Alan Bruce as symbols of quite how extreme our circumstances have become. I hope their actions can motivate people to take stock, take care and remain connected to loved ones. Neither of these men's families or friends had any idea of their intentions. Terry Kaelber, who was Buckel's husband, said that Bruce's act 'did make me think of what David did and also the incredible pain this sort of act causes the people who love them.' One of the most important things we need to do in the face of collective struggle is to remain open with each other and mutually supportive. I know from experience that anger can quite easily be misdirected and end up unintentionally traumatising those we care most deeply about.

The second reason I was reluctant to relate these events in a book about climate change and mental health is that some deny they are suicides at all. One Zen Buddhist priest, a friend of Bruce's, announced that it was 'not suicide' but 'a deeply fearless act of compassion to bring attention to the crisis.'[20] No matter how these acts are interpreted by Buddhist scripture, both men undeniably died by their own hand. Unless you believe in reincarnation, they are gone from this world. It is possible to be simultaneously grateful for their actions, in life and in death, whilst also committing to the belief that there are other transformational ways to influence the climate crisis and protect those we love.

Climate psychologist Lise Van Susteren, who co-founded the US-based Climate Psychiatry Alliance, recently talked to the *New Republic* about climate change and suicide. She acknowledged the complexity of the situation, and of defining causality, then said flatly: 'here's the thing, [on] my dashboard, the lights are all blinking red.'[21]

People working directly on climate change appear to be particularly susceptible to climate despair. Climatologist Judith Curry, amongst others, was talking about 'pre-traumatic stress syndrome' as far back as 2015. She quoted climate scientists speaking in a number of publications as living a 'surreal existence,' suffering from 'obsessive intrusive thoughts,' and being 'professionally depressed'. Today things are even worse for climate scientists and activists. An international project called *Is This How You Feel* set out to collect letters from prominent climate scientists from 2014–15, then followed up with the same respondents five years later to see how their mental state had changed. Professor Katrin Meissner, Director of the Climate Change Research Centre at the University of New South Wales had this to say in 2020:

I feel powerless and, to a certain extent, guilty. I feel like I have failed my duty as a citizen and as a mother because I was not able to communicate the urgency of the situation well enough to trigger meaningful action in time. What we are doing right now is an uncontrolled, risky experiment with the planet we live on.[22]

Climate despair is rife. Not everyone seems to agree about what is going on. I have come across four key orientations when it comes to interpreting what these kinds of experiences could mean:

1. **Cracking up with the climate**. Climate change and society's response to it (or lack of) is a a contributory factor resulting in and exacerbating mental-health issues.

2. **Adapting to the atmosphere**. Climate despair is a rational and healthy response to an apocalyptic situation, but society is unnecessarily pathologizing it.
3. **Chuck it in the bucket**. Climate change is a vessel into which people are pouring their pre-existing mental-health issues, rather than climate change being the cause of anything new.
4. **Same old, same old**. This is just another example of suffering; people feeling bad about climate change isn't special.

Most cases fall into 1. Lots of us are cracking up with the climate. I have sympathy with 2, the idea that we're adapting psychologically, and often this is the case. I also believe there are quite a few examples of 3. Number 4 is bullshit. Let's discard that one. All the others are disquieting. 'Cracking up with the climate' and 'chucking it in the bucket' are both problematic, even if the latter is just climate supplying additional nightmare fuel for those already struggling psychologically. 'Adapting to the atmosphere' is often presented as being essentially fine. It is good, people say, that our minds are respecting the gravity of the climate situation. But if people are being driven into psychological distress because of climate change, calling this 'adaptation' can be a troubling euphemism that masks systemic suffering. R.D. Laing defined insanity as a rational response to an insane world. That doesn't mean the experience of insanity is okay. It doesn't mean the world is okay. It does mean, though, that we are all in a profoundly upended and warped reality – an indictment of civilisational proportions if ever there was one.

There are established neurobiological pathways that link chronic stress, which can start as climate anxiety, to issues from depression to pronounced substance-use disorders.[23] In general, personal breakdowns that lead to depression often come in the wake of break-ups, lack of economic security, grief and/or public humiliation. Each of these constitute a shattering of one's imagined future, as well as a pronounced shift in one's status in relation to the world. Accepting the daunting threat of climate change demands a rupture between one's self and the accepted narratives of modernity. The same goes for our previous assumptions about our life's course. We must break up with the status quo, contend with diminished expectations of security, experience a loss of hope, alongside the risk of humiliation, should we openly communicate our true fears. Research suggests that even people who care deeply about the climate rarely talk about it, often for fear of upsetting people, bringing the mood down or alienating loved ones.[24] Depression can, unsurprisingly, ensue. Some researchers have talked about a 'socially constructed silence' around climate-change emotions. In the long run, this drives up anxiety.[25] Drugs and alcohol offer a poisoned chalice disguised as respite. It's very tempting to give in.

In the UK, NHS prescriptions for anti-depressants have increased by 35 per cent in the last six years.[26] A friend recently commented to me that it seems everybody is depressed nowadays. 'It's just a symptom of living in the end times,' she said, jokingly adding that on a first date it is now pretty common to ask people what prescription meds they are on. We are skilled at masking this stuff, right up until the moment it's no longer possible to function.

For many of us, it is easier to imagine ending our lives than it is to imagine the world being transformed for the better. I attempted suicide because I didn't want to live in this world. That is still the case, but the emphasis has changed. Before, I didn't want to *live* in this world. Now, I refuse to live in *this* world. When people say our way of life is suicidal, I no longer see it as an overreach. Coming to terms with that systemic darkness and trying to dislocate from it to fight for something better inevitably forces a reckoning.

I spoke with Dr Emma Lawrance, author of several important studies on climate change and mental health. She leads mental-health strategy at the Institute for Global Health Innovation as well as the Climate Cares programme, an interdisciplinary project looking to understand and support people's minds in the context of ecological crisis. As an academic, you might expect her to be diagnostic and discrete in her definitions of climate-induced distress. What she said surprised me:

> Once you open these things from the starting point of the climate crisis it links into our understandings of ourselves, of 'mental illness', of society, philosophy, and our economic system. To say 'where do you put a line around what's clinical, what's non-clinical, what's depression and what's sadness,' in response to climate change – these questions are wider than eco-anxiety. They're about how we think about mental health in society in general.
>
> Of course the diagnostic criteria for mental illnesses can be useful, and it's important to acknowledge the enormous life impact such illnesses can have – as I myself have experienced. Indeed, the climate crisis is compounding the risks faced by people with mental illnesses and disrupting access to care. But it's also vital to understand that our biology and psychology (our minds!) interact with the context we are in. Grappling with the climate crisis – the traumas, losses, but also opportunities to foster a world better for our mental health – really brings that to light.

At the time of writing we are at about 1.1 degrees Celsius of heating.[27] The World Meteorological Association predicts that there is a 50:50 chance that one out of the next five years will be 1.5 degrees Celsius warmer than pre-industrial temperatures.[28]. Back in 2018 the Intergovernmental Panel on Climate Change (IPCC) was widely quoted as saying we had twelve years to avoid 1.5 degrees Celsius of heating. The same report claimed that for emissions to be on the right trajectory, reaching the necessary 45 per cent global cut by 2030, international emissions had to peak in the year 2020.

Astonishingly, and for the first time in living memory, global emissions did drop in 2020. They dropped by 2.3 billion tonnes overall, the largest reduction in emissions since the Industrial Revolution – a global reduction of around 6 per cent.[29] Half of this was the result of reduced transport use, particularly road transport and aviation.[30] Over the course of the year, emissions from flying fell by 48 per cent.[31] Energy demand dropped too.

Of course, 2020 was the year the pandemic broke out. The emissions reductions can be mapped pretty neatly onto the international proliferation of policies that kept people in their homes to avoid the spread of coronavirus. By December 2020, emissions were already back up to 2 per cent above the same month the previous year.[32] In 2021 we collectively emitted more carbon dioxide, methane, nitrous oxide and other climate warming gases than at any point in history.[33]

This is all very hard to take in. Our civilisation has fundamentally disordered the climate. The climate is now disordering our minds. It is scary that we are only just starting to look at it with any seriousness. The psychological threat of civilisational collapse is already imperilling millions. Many will never receive the care and support they need, a situation that threatens to plunge them and their families into lifelong existential struggles. Their suffering is not understood, rarely shared and exists mostly in the deep, dark and lonely recesses of minds scattered across the globe.

2

The ecology of mind: our managed madness

'Sensations are the building blocks of our experiences – meaning, at the base of every internal experience is sensation. Understanding sensations as a foundational language, we can then feel emotions. . . . Emotions are deeply meaningful to us and can also act as guides to our commitments, connections, healing, and growth.'

Staci K. Haines, *The Politics of Trauma,* 2019

THE PHILOSOPHER TIMOTHY MORTON'S term 'hyperobject' describes a phenomenon so large that it is beyond the realm of human comprehension. Prior to the pandemic, Morton's favourite hyperobject was climate change. The reason I bring it up here is that we are all in a position where something we cannot possibly understand in its entirety is nonetheless making us unwell. In an early paper on the concept, Morton writes:

Inside the belly of the whale that is global warming, it's oppressive and hot and there's no 'away' anymore. And it's profoundly regressing: a toxic intrauterine experience, on top of which we must assume responsibility for it. And what neonatal or prenatal infant should be responsible for her mother's existence? Global warming is in the uncanny valley, as far as hyperobjects go. Maybe a black hole, despite its terrifying horror, is so far away and so wondrous and so fatal (we would simply cease to exist anywhere near it) that we marvel at it, rather than try to avoid thinking about it or feel grief about it. The much smaller, much more immediately dangerous hole that we're in (inside the hyperobject global warming) is profoundly disturbing, especially because we created it.[1]

This last point, our having 'created it', is intimately tied to guilt. In recent years it has been established that excessive guilt in childhood can influence the development of the mind.[2] The mind, I would argue, is a hyperobject of a different kind, one so ethereal and non-descript that its rules elude us (if there are any). In a series of shocking studies, researchers found that excessive guilt in children can have a physical impact on the development of the brain. One study carried out multiple assessments of children over the course of ten years and found that pathological guilt reduced the size of the anterior insula area of the brain. Another found cortical thinning in the brains of pre-schoolers with maladaptive guilt. These, in turn, are strong predictors of depression. Even toddlers under three years old who displayed symptoms of pathological guilt were found to be ten times more likely to develop major depressive disorder by the age of five.

I have two observations to make about these findings. First, I don't want you to think that depression is simply a neurologically pre-determined phenomenon. Like pretty much all mental-health issues, depression is something that is *primed for.* Genetics and physical health play an important role, but so do environmental factors, such as upbringing, social status, identity, the lived environment, economic security, substance use, trauma and, importantly for our purposes, the state of the world. No one is destined to be depressed, just as no one is destined to be happy, even if the cards we are dealt are stacked a certain way.

Second, it makes me livid that people are made to carry so much of the burden of this catastrophe, in stark contrast to how responsible they actually are. I was born in 1992 and it was terrifying for me to find out that around half of the emissions spewed into the atmosphere since 1751 have been produced during my lifetime. It is easy to feel responsible for some of that, especially alongside all the marketing and PR promoting consumer-facing ways to cut our emissions. It is also easy to feel horrified that during this brief flash of time I have been alive, the species I belong to has managed to upend the global climate system, bringing atmospheric carbon to a level not seen in 3 million years.

I am one of 7.753 billion people on the planet now. Over my lifetime we have emitted around 50 per cent of the carbon released into the atmosphere since the Industrial Revolution. If I were to take an average[3] then I would be personally responsible for 0.0000000064491165 per cent of all historical emissions.

To express that in a format we can relate to, if we converted all historical emissions into the distance between the earth and the moon – approximately 350,000 kilometres – the average person's emissions over the last twenty-five years would get us 2.5 centimetres of the way there. That is about half the height you would gain by standing on your tiptoes. Excessive guilt, often a driver of serious mental-health issues, is inappropriate and maladaptive. Any climate guilt at all, I would argue, is almost irrelevant. It hurts the sufferer, and by extension limits their capacity to enrich other people's lives. Over half of young people globally report troubling feelings of guilt. It is astonishing that fossil-fuel companies have so deftly shifted the blame onto individual consumers. We should never forget that it was BP, after all, that first developed the concept of the individual carbon footprint.[4] Honestly, though, I'm even more blown away by our ability to feel complicit in something so utterly beyond our control and comprehension.

I don't want people to feel guilty about the guilt, though. Meta-guilt never helped anyone. Instead, we should let ourselves feel angry about the sleight of hand played on us, the ways in which we have been tricked into self-flagellating. Most contemporary treatments for mental-health issues focus heavily on the individual in a similar way. This can be punitive victim blaming, with recovery models that put the onus squarely on the shoulders of the sufferer, rather than more communal and systemic responses. The biomedical approach so beloved by the Western psychiatric profession also acts as a sidekick to extractivist capitalism, pushing people back into the very systems that made them sick. As the psychologist Sanah Asan, referring to the individualised model of health, puts it: 'what's most devastating about this myth is that the problem and the solution are positioned in the person, distracting us from the environments that *cause* our distress.'[5]

The biomedical model of the mind is prominent in Western psychiatry. It tends to view our psyches as machines, like computers that 'glitch' when we have mental-health issues. Most people are familiar with the phrase 'chemical imbalance', in relation to depression, psychosis and the like. What many practitioners believe this necessitates is a chemical or physical input – like medication, exercise or, in extreme cases, electro-convulsive therapy (ECT) – to reboot or rebalance the system and return it to normal functioning. This model is a piece of crap. We shouldn't think of our minds as isolated computers. The model is unscientific and, importantly, at odds with our lived experience. We are not as atomised as we have been led to believe. We can't let the cult of individualism steal its way inside our heads.

Much better models, and ones that are gaining traction with scientists and practitioners worldwide, are those that see our minds as relational ecosystems, intertwined with external psycho-social realities. We are embedded in social environments. We are embedded in ecology. The boundaries between the self, the social and our wider environment are breathable. They are porous. We might be used to seeing our minds as a black box, a vivisected unit as disembodied and dismembered as our hollowed-out modern idea of community. In reality, we are inextricably 'braided' into the world, to use Jayne Engle's phrase. The world moves through us as we move through it. Very often, psychological imbalances are a reaction to imbalances in our social and environmental realities. No man is an island. That's a good thing. A lot of islands are going to be under water soon.

Stress and carbon dioxide

All of us can be pushed towards mental disarray by circumstance. Despite what some might think, no one is immune. Climate change is the quintessential 'everything crisis.' It is going to hit our heads using lots of different pathways. The 'allostatic load' model is one helpful way of understanding this.[6] According to the model, there is a certain amount of stress that a human can take before the body either falls apart in some way, or overcompensates and creates maladaptive side-effects. Chronically high or wildly fluctuating levels of stress can seriously impact the nervous system and the endocrine (i.e. hormonal) system. There are sustained links between allostatic overload and the onset, duration and severity of many mental-health issues, including anxiety, depression, post-traumatic stress disorder (PTSD) and psychotic disorders.[7]

How much stress we can take is different for everyone. We each respond to stress differently. We each have a collection of risk factors such as historical traumas, existing mental-health issues or difficult relationships, plus lifestyle factors like diet, fitness and drug and alcohol use. We also have our own individual support factors, such as the stability of housing, our financial circumstances, access to healthcare, and our webs of meaningful relationships. Regardless of how much support we have, there is always a limit to how much any one individual can take.

We can think of it in a similar way to the climate system. Imagine stress is carbon dioxide. Our risk factors can be thought of as the amount of carbon dioxide already in the atmosphere. Our support factors are the oceans, the trees and the plants that absorb the carbon dioxide. If there is more carbon going into the atmosphere than is being drawn out, then carbon dioxide levels will go up. If the amount of carbon in the atmosphere exceeds a certain amount, the risk of destabilising the equilibrium of the climate increases. This is when we see outbreaks of disaster.

Not all mental-health issues are caused by stress, but it is a useful framing nonetheless, especially given that stress can be defined in wide-ranging yet precise ways. According to the physician and author Gabor Maté, there are three factors that universally lead to stress. These are 'uncertainty,' 'lack of information' and 'the loss of control.'[8] We directly experience these three factors in our lives every day. We also experience them in relation to the structure of the world and our narratives about the future. In terms of the climate crisis, we live them in the extreme, even from a distance. We lack 'certainty' about climate change. We lack 'information' about climate change. We lack 'control' over climate change.

The person's silhouette is adapted from Sid G Hedges' 1925 *Swimming, diving & life-saving*. The world map is the AuthaGraph projection, an innovative project by Hajime Narukawa aiming to accurately and equitably display land mass and redress cartographic and geopolitical distortions.

The fear system, principally based around the amygdala, is responsible for anxieties relating to external threats. The panic system, rooted in completely different neural structures, is responsible for anxieties connected to loss. As Joseph Dodds points out in the *British Journal of Psychiatry*, these two systems map neatly onto both of the categories of anxiety defined by the psychoanalyst Melanie Klein, as well as two principal forms of climate anxiety. Klein talked about 'paranoid-schizoid anxieties', which, like the brain's fear system, focus on external threat. Dodds convincingly argues that this is very similar to climate anxiety's more 'apocalyptic' terrors: fears of death, annihilation and extinction. Klein's second class of anxieties, the 'depressive anxieties', are more concerned with loss. This aligns with Dodds' second type of climate anxiety: feelings of grief, dependency, loss and guilt, each connected to losses that have already happened or are suspected to arrive. Across the often-competing disciplinary areas of neuroscience, psycho-analysis and psychology, here there seems to be alignment.

What might evolutionarily have been healthy responses to threat can now very often be unhelpful, living as we do in societies monu-mentally different from those we evolved in, despite having similar physiologies. The autonomic stress response, also known as 'fight, flight, freeze,' or F^3, was once a fairly effective way of escaping predators. When the threat is as abstract as climate change, it can be less adaptive. With climate despair and climate trauma, the 'fight' is very often turned inwards. It is played out as persistent conflicts in the mind, our various selves battling it out or otherwise conspiring together to deviously ferret out our most sensitive vulner-abilities and use them to cause as much damage as possible.

The 'flight' response often takes the form of escapism. This can be physical and/or psychological. It can involve escaping into fantasies of running off into the hills, finding comfort in an insulated and involuntary dream-like state where nothing matters, or diving head first into hedonistic nihilism. This can all be difficult for others to detect, as pursuits like addiction are often secretive, and more alluring because of their very secrecy and impermissibility. In other cases, a flight can be more obvious. As we have seen since Covid, many people are barricading them-selves into their bedrooms, obsessively playing videogames to the extent that they effectively spend most of their time in an alternate reality. A flight response usually incorporates strong elements of denial, or at least a necessary psychological 'splitting' where the sufferer lives in a dual reality, unable to unify contra-dictory versions of the truth.

Writ large, there's no way for us to run away from climate change. Obviously, despite the efforts of Elon Musk, Jeff Bezos and Richard Branson to colonise the solar system, earth is the only life-sustaining planet we know of in the entire universe. Its homeostatic controls are now being corrupted by civilisation. Even billionaires with apocalypse bunkers have a questionable level of sustained safety should the worst climate projections come to pass. The rest of us have nowhere to go, except by dislocating ourselves psychologically. We can see the pursuit of alternate and modified realities – Meta, Neuralink, etc – as extreme profit-based escapisms offered by the high priests of tech. It's understandable. If (big if) most of our reality is experienced inside our minds, a mental exodus can be an effective short-term strategy for avoiding catastrophe in the material world. It can be terrifying to be thrown into this state, but there is an entirely comprehensible survival logic to it.

The 'freeze' response is a relatively modern addition to the 'fight or flight' model. In a climate context, some call it 'eco-paralysis'. For most, this plays out as an experience of overwhelm. Many are haunted, if not emotionally and psychologically kneecapped, by the unravelling of the climate system, the dissolution of our civilisation's narratives of progress, the relentless crises we're assaulted by and the expectation that they will only get more frequent and more extreme. Seizing up is not a surprising outcome. As the old saying goes: 'If you can't see where you're going, stand still until the fog clears, otherwise you might end up walking off a cliff.'

Freezing can be maladaptive, though. Failing to perform daily tasks can rob people of their autonomy, personality and even access to the means of survival. I can tell you from personal experience, having lost jobs because I couldn't physically get there, that whilst freezing up might make a certain amount of evolutionary sense, it isn't conducive to survival under neoliberal capitalism.

I have been heavily dissociated for several years. Initially this was terrifying. I was constantly questioning whether I existed, whether the world was 'real', and trapped in a dizzying maze of abstract thought patterns. It simultaneously felt like I was in a dream, or acting in a film where every word I spoke was already pre-ordained. It wasn't something I could control. It was as if I'd been spiked. For a time I turned to alcohol (another altered state but at least one I could tell myself I'd chosen voluntarily). When I discovered that dissociation was an evolutionary survival mechanism, something shifted. If you are being eaten by a tiger, it might help if your mind pulls you out of yourself for a minute. 'Oh look,' you say. 'There's a tiger.' You depersonalise and derealise. It's an effort to cope. You may even be able to escape, slipping away calmly in the same way your identity has slipped away from your psyche. Dissociation, for me, is protective. It wraps me in an uncanny cocoon where everything is muted and strange.

To be clear, there are still elements of my mental health that fuck me up. Depression is a dick. Anxiety can put a serious spanner in the works. PTSD sometimes confines me to my room for days on end. It can make relationships difficult and get in the way of survival. Psychologists sometimes talk about elements of our mental health as either adaptive or maladaptive. Eco-anxiety, for instance, is described as 'adaptive' if it helps people to align themselves with the world gainfully. A recent paper in the *Journal of Climate Change and Health* found that sufferers of 'eco-anxiety' who allow themselves to get angry rather than succumbing to depression have a strong drive towards both climate action and better wellbeing. As the authors put it, 'eco-anger predicted better mental-health outcomes'.[9] Feeling terrible about the climate crisis is healthy and useful, in other words, if it can be catalysed into meaningful, connected action. The professor and author Donna Haraway, who likes toying with the English language, breaks the word 'responsibility' into the words 'response' and 'ability.' For her, and I agree with this, responsibility should be about agency, action and a way of expressing and experiencing our interconnection with others.

Mental tipping points
'Maladaptive' eco-anxiety, according to the American Psychological Association, can lead to panic attacks, irritability, loss of appetite,

weakness and insomnia. Other sources even report sufferers feeling physically suffocated.[10] In the *Journal of Anxiety Disorders*, Susan Clayton argues that maladaptive eco-anxiety arises when 'the sensitivity to potential problems – differences between what is expected and what is encountered – is too great, triggering an emotional response and rumination that inhibit resolution of the anxiety.'[11] Like the destabilisation of ecosystems and the climate system as a whole, our minds can reach a tipping point where the problems get worse and worse. The strategies we normally rely on to rebalance us fail. This is not something we have much control over.

Often, then, 'good' mental health is about coping. With mental health being such a complex and nebulous area, with serious suffering sometimes normalised, whilst at other times utterly valid and healthy states of mind are pathologised, whether or not we're 'coping' mentally is an important way of deciding whether we should be concerned. The WHO defines mental health as 'a state of well-being in which every individual realizes his or her own potential, can *cope* with the stresses of life, can work productively and fruitfully and is able to make a contribution to her or his community'.

The use of the self-determining word 'cope' is great. It might seem vague but 'coping' is an effective shorthand for an expansive collection of factors, a shorthand that is still meaningful despite its simplicity. It is much better than being deemed 'ill' or 'well'. Coping combines concepts of resilience, capacity, risk factors and triggers, as well as respecting people's priorities even if they're internal and subjective. This multifaceted concept, coping, leaves room for us to describe things with the necessary nuance and subtlety. It is more humane than mere pathologising.

Psychiatry and psychology are never politically neutral. Downplaying the social and ecological drivers of our inner worlds has the effect of forcing people back into the systems that made them unable to cope in the first place. It feeds sufferers back into neoliberalism, making them tools rather than people. Even the WHO definition above mentions the need for people to 'work productively' in order to be deemed 'well'. In other words, it demands that we squeeze ourselves into a narrow definition of productivity, a normalcy that is largely responsible for the climate crisis and the mental-health epidemic in the first place.

This is the result of a millennia-long process of extractivism, an output of our disconnection from and domination over nature and each other. Murray Bookchin talks about this as our being 'disembedded' from the natural processes that keep us well. Extractivism turns people into units of productive output 'just like,' Bookchin writes, 'the tools and machines [people] create. [We], in turn, are subject to the same forms of coordination, rationalisation, and control that society tries to impose on nature and inanimate technical instruments.' This inescapably warps our minds, by warping our relationships with each other and decimating the ecological settings we evolved in and rely on. 'Self-repression,' Bookchin continues, 'and social repression form the indispensable counterpoint to [the] personal emancipation and social emancipation' that technical progress promises.[12]

Models that individualise and pathologise people's mental-health issues feed this system, turning people into atomised things. They overemphasise medication, one-to-one therapies and personal behavioural adaptations that expect patients to change to accommodate the world around them, rather than considering how the world could change around us. This negates the social and ecological drivers of mental-health issues. It also negates social and ecological models of recovery. Leaving the wider world outside the therapy room is no longer an option. Some psychologists are lamenting the fact that they 'feel unequipped to handle' climate-induced mental-health issues.[13]

This has long been a problem, with poverty, oppression and cultural genocide long separating communities from the ability to cope with life. Liberation psychology, somatic healing and community models of mental health have operated in marginalised spaces for decades. Now we are seeing models of climate-relevant psychology being born. The Good Grief Network, for instance, convenes 10-step-style peer support groups around the world.

The Climate Psychology Alliance offers spaces and therapies held by climate-aware practitioners, helped by the fact that there are now accredited qualifications in climate psychology.[14] A team of researchers, designers and affected individuals at Climate Cares are developing practical tools for coping with what they call 'eco-emotions', whilst outfits like Force of Nature and Tipping Point are co-designing ways for distress to be transformed into climate action and connection as modes of recovery. These extend beyond recommendations to go for a walk in the woods. They are invitations to dig deeper into what is making us mad, how to build communities of care and actively engage in the fight for a better future. There is a tendency in the biomedical and research community to publish vast tracts, sometimes thousands of pages, most of which conclude with words to the effect of: 'this is bad'.[15] The next question – 'what are we going to do about it?' – often goes unasked, let alone unanswered. The people we will meet in the coming chapters refuse to be held back by such barriers.

For someone to have a mental-health issue, be it depression or psychosis, the person needs to fit certain medical criteria. These are presented as clear-cut demarcations. They are not. The ecosystems of the human mind are so complex, amorphous and ultimately unknowable that slapping a label on anyone's head is as reductive as defining the moon as just a white circle. According to Donald Hoffman, our understanding of the mind today is equivalent to Galileo's grasp of astronomy 500 years ago. We know there are half as many neurons in each person's head as there are stars in our galaxy, but we don't know how our conscious experience of the universe emerges from that unfathomably complex system. We are, as Bill Hicks once quipped, the universe experiencing itself subjectively. Beyond that kind of poetry we don't have much.

Anil Seth, a professor of cognitive and computational neuroscience, talks about consciousness as a 'controlled hallucination'. There are so many stimuli being relentlessly fired at us that we must fabricate structures that are intelligible to us in order to make any sense of the world. The colour red does not 'exist' as such, at least not outside the mind. It is something we have invented as shorthand for a certain intensity of photons hitting our pupils. Emmanuel Kant highlighted a similar need for combining the empirical and the rational to make sense of anything we perceive. My favourite example of his is the simple act of counting. We can only count things out in the world if we have mental structures that reductively categorise endlessly complex and different structures of atoms as chairs or birds or stars. Really, we make them chairs or birds or stars by fabricating their definitions. A friend of mine described this to me recently as 'cleaving certainty from the infinite'. I call this exercise in comprehending chaos our 'managed madness'. We are all mad. We have to be to survive. Sometimes we manage our own madness. At other times, it is managed against our will by people more powerful than us.

Conventionally, the criteria that define madness or 'mental illness' are largely based on deviations from socially permitted behaviours and perspectives on reality. Interventions are often decided based on the extent to which these deviations from the subjective norm bump up against an implicit social contract. This might sound like a sensible approach. But if what is deemed 'acceptable' is defined by an unjust authority, then diversity and dissent become pathologised, the diagnosis of madness is weaponised, and 'health' provides the lockbox to put the crazies in. This gives an ideological justification for dehumanising us and delegitimising our concerns.

The parameters of sanity and moral acceptability are continually moulded and remoulded. This can incarcerate dissenters, rebels and visionaries (delusional or otherwise) who challenge the dominant culture, often trailblazers who innovate, invent and open doors to other worlds. It is only recently that support for those of us with mental-health issues has become a rhetorical priority. For most of history, psychiatry has provided the powerful with excuses for punishment and persecution. It has evolved in the last few centuries and is now more compassionate, generally. But it is not an entirely different beast. Psychiatry is a disciplinary practice and its oppressive history survives in much of its DNA: refashioning the mad into 'productive' citizens (where possible), sedating and infantilising divergence (where not), whilst defanging resistance, pacifying rebellion and invalidating visions of alternative ways of life.

Deeming minorities 'mad'

When diagnoses are the result of people showing the effects of oppression, psychiatry is a conveniently authoritative, epistemological framework to neuter systemic critiques from oppressed majorities. In 1982, at the tenth international annual conference on human rights and psychiatric oppression, a group of 'psychiatric survivors' agreed a statement of principles that outlines this effectively. They wrote:

> We oppose the psychiatric system because it feeds on the poor and powerless, the elderly, women, children, sexual minorities, people of colour and ethnic groups . . . because it invalidates the real needs of poor people by offering social welfare under the guise of psychiatric 'care and treatment', [and] because it creates a stigmatized class of society which is easily oppressed and controlled.

This statement was penned less than a decade after the American Psychological Association (APA) took homosexuality off its list of diagnosable mental illnesses.[16] It took the WHO until 2018 to stop classifying being transgender as a psychiatric condition. The psychiatric handbook used in the US and much of the rest of the world[17] still lists gender dysphoria as a mental disorder. In the UK, a psychiatric diagnosis is required as part of a trans person's application to change their birth certificate.[18] Being trans and non-binary is not a disorder. The way these communities are treated – including social isolation, abuse and structural discrimination – can precipitate huge mental distress. Some three-quarters of US trans and non-binary youth have clinical depression. According to one big study, a horrifyingly high figure of 23 per cent attempted suicide in the last year. Those granted access to treatment have far better mental health.[19] Fascists and populist demagogues are using the fabricated 'trans debate' to cling to power, a debate that hinges largely on the designation of trans people as crazy or deranged. When the former Scottish First Minister Nicola Sturgeon changed the law to allow gender self-determination, UK Prime Minister Rishi Sunak vetoed the change. It was the UK's first ever application of the 1998 Scotland Act, weaponised to undermine Scottish sovereignty. Whilst people like Meloni, Putin and Trump would surely continue deriding gender diversity whatever happens, designating gender dysphoria as a mental-health condition justifies their arguments, a justification cloaked in the cultural authority of medicine.

Women have been accused of hysteria – a fictitious mental disorder characterised by excessive emotions and neuroticism – as far back as Ancient Greece. The Ancient Greeks thought that hysteria was caused in women by the womb physically migrating around the body. The Victorians put it down to a weak nervous system. For generations, women who refused to obey patriarchal standards could be punished on pseudo-medical grounds, an avenue for persecution with less cultural blowback than bald violence. In *Caliban and the Witch,* Silvia Federici relates stories of the European witch trials of the 13th to 18th centuries. They were, Federici argues, essentially a way to control women and their bodies and to stamp out female perspectives on the universe[20]; deeming them insane and demonic meant that they were classified as antithetical to progress and civilisation. Burning and drowning these women was rationalised by warped definitions of acceptable psychologies. As Federici writes, this 'destroyed a universe of practices, beliefs, and social subjects whose existence was incompatible with the capitalist work discipline'.

Psychological dehumanisation was sharpened to a point during the colonial project. Much like accusations of demon worship subjected European pagans, colonisers commodified the populations of the Americas, Africa, Australasia and Asia by positioning the beliefs and practices of the colonised as evidence of their sub-status. They didn't conform to extractivist norms, and so they were objectifiable. Being constantly drilled on the superiority of whiteness put immense, unprecedented strain on the minds of the colonised. Frantz Fanon argued in *The Wretched of the Earth* that one of colonialism's explicit tactics was to deprive colonised people of the very notion of a 'self'.

British medics regularly pathologised Africans who tried to revolt against Empire. Revolutionaries in Nigeria, doctors claimed, were apparently suffering from 'mass hysteria'. It was utterly fictitious but a highly effective method to explain away the existence of so-called 'rebellious types'. It also helped bring about the reclassification of uprisings as the machinations of the mentally ill rather than genuine expressions of justified discontent – instances of sick troublemakers spreading their illnesses in a way that led to 'psychic epidemics'.[21]

Later, psychiatry proved helpful in mollifying opposition to slavery. 'Drapetomania' was the mental disorder coined for why slaves might want to run away.[22] Blackness itself was even classed as a mental illness for a long time.[23] Relics of this pathologisation survive in much of our culture, nurtured by the dominant white race even in the very maintenance of 'whiteness' as a concept. 'To truly savour the satisfactions of race,' write Mary Schmidt Campbell and Lucille Jewel, 'whiteness must instead camouflage "the social as the natural" by shoving its identity as the author of its racial mythologies far below the surface.'[24] As the neuroscientist Araceli Camargo said to me, it is whiteness itself that should be considered the 'cognitive virus,' not the symptoms of the oppression that whiteness forces onto others.

Industrial civilisation methodically undermines many of the means to cope with the maladies it exacerbates, not just on an individual or community level, but at the level of social transformation. Wilful misdiagnoses of mental-health issues have been fundamental in discrediting and imprisoning political activists – sometimes called 'punitive psychiatry'. The US government sectioned many civil-rights activists and schizophrenia has been used as a form of racialised social control against the Black community. Declassified documents now show that the FBI diagnosed Malcolm X with 'pre-psychotic paranoid schizophrenia'.[25] Ironically, J. Edgar Hoover, then head of the FBI, was a textbook paranoiac (and he definitely heard lots of voices). As recently as the 1970s and 1980s, one third of the population of the Soviet Union's psychiatric hospitals were actually political prisoners.[26]

Today, the Chinese Communist Party has reportedly sent one hundred political dissidents to psychiatric units to bypass legal accountability.[27] Europeans diagnosed with mental-health issues who have committed a crime are still frequently imprisoned for significantly longer than their sentences. Their psychiatric diagnoses mean a psychiatrist can deem them unfit to re-enter society. This can be difficult if not impossible to challenge.

More colloquially, radical progressives are often derided as the 'loony left.' It is hard to take this lightly given the historical roots of delegitimising dissent using psychiatry and the continued persecution of many due to ailing and failing medical systems. Words like 'mad', 'insane' and 'crazy' typically have negative connotations. But they are also linked to ideas of creativity, foolishness, brilliance, randomness and the unknown. Irrationality can be a surprisingly destabilising and generative force. For a mechanistic culture built on vivisecting the world into smaller and smaller discrete units, you can see why that's a threat.

Joan of Arc was burnt at the stake because she had visions and claimed to be blessed by direct contact with God. She is now venerated as a saint by the church that put her to death. My grandmother, whilst not officially a saint, was violently punished for her unfeminine wilfulness. She suffered from depression, but psychiatric brutality was often wielded as a threat. When she tried to leave one husband and take her son with her, she was threatened with involuntary sectioning. She gave in. A later partner did manage to have her sectioned and she was forcibly given ECT. My great-grandmother was described by the family as 'feeble' and 'of nervous disposition'. Truly she was depressed, but she was also resistant to being a quiescent wife. This was turned into a pathology and so she, like my grandmother, was forced to have ECT. She later said it diminished her capacity to complain or act on her feelings. The feelings themselves didn't change at all. Not so long ago I was coerced into having ECT against my will. Thankfully I had the benefit of general anaesthesia. I had completed a series of treatments, consensually, and it had helped with my suicidality. My psychiatrist at the time wanted me to continue having it monthly, indefinitely, as a preventative. The induced seizures had already seriously degraded my memory. When I resisted, he said I could withdraw my consent, but that he was still going to prescribe it. If I didn't attend, he told me, I'd be labelled as a non-compliant patient. I would therefore lose priority should I ever request it again, putting my access to emergency treatment at risk. I went, fearful and powerless. I consented, against my will.

There are so many other ways to respond to mental divergence than to cram people back into a socially acceptable mould. We have been systematically deskilled at looking after one another, but there are generations of experience to draw from, people applying reconnection and equity to the relational ecosystems of our minds. Madness can be horrible to experience, both internally and for those around us, human and non-human. But it is only by relating more creatively to madness that we can ever hope to stop treating the mind as an extraction site. We have to learn to live with difference and unlock the visionary potential of other states of being.

Follow the yellow brick road
In the early days of my recovery I occasionally thought of Dorothy on the yellow brick road. I was partial to fantasising about alternative histories and technicolour versions of the future I wasn't sure I had ever really believed in. At the end of her journey, it was undeniably traumatic for Dorothy to see behind the curtain and understand that the great wizard was just an old man with some impressive tech-facilitated illusions. It was

similar for me at the UN climate talks, realising that these (mostly) old men were unlikely to do anything meaningful about the crisis. It is the same for many when they first encounter the reality of climate disaster. But pulling that curtain back was vital for Dorothy to see the reality of her and her friends' situation, if only to help resolve their struggles.

In the lesser-known 1985 Disney sequel, *Return to Oz*, Dorothy is back in Kansas, but her family believe that her obsession with the land of Oz marks her out as insane. She is institutionalised and is about to be given electro-convulsive therapy against her will, when lightning strikes the psychiatric ward, sets fire to the building and Dorothy is whisked away to Oz again. There she finds a world torn apart by evil forces, inhabitants turned to stone and the yellow brick road (to recovery) fragmented. She makes new friends, one of whom is a robot presciently named Tik-Tok,[28] and is forced to fight a battle on a much grander scale than before. The entire fate of Oz is at risk, but with character-istic guile, fortitude and occasional bouts of despair, Dorothy and her friends eventually restore the Emerald City to its previous splendour. She leaves with the aid of the newly appointed Queen of Oz. Back in Kansas, the psychiatric ward has burned to the ground. The fire killed only the sadistic chief doctor. Once at home, Dorothy sees her friends from Oz through a window in her bedroom mirror. She is about to call her aunt and prove her sanity, but her friends tell her to keep Oz to herself. She no longer has anything to prove.

This darker, more elaborate and true-to-life (or at least true-to-trauma) telling was, unsurprisingly, less popular with audiences than the original. In fact, at the box office *Return to Oz* made back less than 40 per cent of the money spent on its production. It may have been darker and spookier than intended, given Disney's family-friendly orientation in the 1980s. Perhaps that was the cause of its unpopularity. The gritty underbelly of mental-health conditions, even fantastical and colourful dressings up of these processes of distress, are hard to look at for too long. That the film was ever made is rather astonishing. Nonetheless, it exists, and is a powerful allegory for the otherworldly feeling of unreality that is embedded in many mental-health conditions, the often harmful individualised treatments offered to those suffering, as well as the need for sustained collective action as a means of recovery.

It is also important to acknowledge that Dorothy emerged from her experiences changed. She didn't struggle to erase what had happened, nor question the truth of her own lived experience. She came out more mature, more in tune with herself and more empowered. As the surgeon Atul Gawande asked in his Reith Lecture, why would we 'waste the experience' of illness, rather than learn from it. In other words, why move backwards, rather than forwards?[29]

We can learn a lot from Dorothy. Responding to the crisis at the root of climate despair in any meaningful way – bearing in mind that we are individuals within a system that appears to be hell-bent on its own annihilation – is extremely hard. We are already incredibly disconnected from one another and it feels like the tide is pulling everything in the wrong direction. Swimming against it would merely increase our levels of exhaustion, despondency and hardship. It's difficult enough to survive as it is.

Expecting someone who is paralysed by fear of ecological collapse to systematically reorient their lives and attempt to avert climate chaos may seem like asking someone who has just come out of a coma to help balance the finances of the hospital they just woke up in. But, as Dr Janet Lewis recently put it in the *Chicago Tribune*: 'The goal is not to get rid of [climate despair]. The goal is to transform it into what is bearable and useful and motivating.'[30] Recovery, whether or not the notion of a 'cure' is appropriate, is a long and winding road whether it's made of yellow brick or not. It can lead us to new versions of ourselves. It is a process usually scattered with moments of revelation, one that takes time but can lead to newfound connection, purpose, meaning, belonging and mutual support.

Normally the question is: how do we get this person back to normal? Normal, of course, is itself defined by the system.

What, then, do we do if it is the system itself that's making us sick? And why is it making us sicker if we try to change it? Here, as with many historical and contemporary struggles for transformative justice, the punitive dominant culture is exposed for what it is. A more systemic and genuinely curative response to so many of us going mad demands a deep dive into the heart of extractive capitalism, into how we got here as a civilisation, and our conception of what it means to be psychologically 'well'. The processes linking climate change and mental health are deeply complex. If we pull at any of the loose threads currently explaining mental health or climate change, definitions of 'illness' begin to blur and alternative perspectives of reality take on a potency and promise. The very fabric of our culture begins to unravel. Thank God.

3

Climate despair,
climate trauma

'It's a lot like the 20th (century) except everybody's afraid.'
Lisa Simpson describing the 21st century;
The Simpsons, S15 E1

THOSE OF US WITH mental-health issues are often branded as being in our own world. We may seem distant, unaware of and unattached to our surroundings, be they people, objects or events. We might instead have a vacant or overzealous otherworldliness. Either way, it appears to leave us unreachable. Paradoxically, though, being in our own world can actually be a result and manifestation of being more connected to the outside world than others, rather than less. In the context of climate change, it may be fairer to describe those who fail to develop symptoms despite the widespread destruction of the ecosphere as being in their own, separate, monoculturally human world, inattentive to the lived experience of billions of other human and non-human beings. They are, in other words, unaffected by looming existential catastrophe. At a certain point, the layers and layers of insulation, made up of civilisational narratives, perceptual blindness and physical distance, dislocate many people from the lived reality of climate chaos. Those whose psyches buckle are much more in tune with material reality, and yet they are often the ones deemed mad. This labelling and pathologising is itself a defence mechanism employed by the civilised to subjugate those whose minds stray from accepted norms. But blessed are the cracked, for they let in the light.

It would seem that anxiety is the mental-health condition designated as the quintessential climate disorder. This makes a certain amount of sense, given that climate change is a pervasive crisis irreducibly characterised by uncertainty, risk and exponentially rising precariousness. The individual experience of anxiety could be viewed as an internalisation of these new global fragilities, responding with excessive levels of worry, sleeplessness, hypervigilance, accentuated fear and debilitating panic.

But climate change can easily be linked to a host of other mental-health conditions. Instability and unpredictability, backed by the threat of severe danger, could just as easily describe the perfect environmental conditions for the onset of post-traumatic stress disorder, or even 'pre-traumatic stress disorder'. Perceived randomness is key to PTSD. In cases of intimate partner violence, for instance, research suggests that the 'unpredictability' of violence is more closely linked to survivors' mental-health outcomes than the frequency and severity of the violence itself.[1] On a global scale, we are living in an unsafe environment. The 'eco' in ecology, after all, comes from the Greek word '*oikos*', or 'household'. Greta Thunberg is right: our house *is* on fire.

We are living in a collective state of unreality, so dissociation and dislocation from reality are similarly apt. This isn't just about denial, nor the fantastically impossible futures of business as usual plus clean tech. We must still survive day to day and persevere with normalcy against a backdrop of impending doom, necessitating a fragmentation of self. This often happens along a fault line of public and private lives, internal and external realities. We struggle with competing and mutually exclusive versions of the future, encounter moral mazes when it comes to personal choices about diet and political activism to whether or not to have children, each of which must be filtered appropriately and repackaged for anyone we interact with. It's not hard to see how this could make some people worry about having split personalities, experience depersonalisation (the feeling that we might not exist) and derealisation (the feeling the world outside might not exist).

The advent of social media and the near ubiquity of screens only accentuates these risks. Diving into these spaces rarely helps people to ground themselves in the 'real'. The alternate realities promised by neoliberal tech utopians are iconic examples of attempts to implement psychotic delusions on a global scale. Back in our everyday world, political and economic reactions to climate change are unhooked from any notion of their real consequences. The post-modern shadowboxing politics of recent despots, from Trump and Johnson to Bolsonaro, Meloni, Putin and Modi is largely fuelled by harking back to a golden age that never existed. This is not so different from the transformed reminiscences of patients with dementia, or the warped ruminations of the psychotic and the depressive.

These socially constructed memories of an imaginary ideal, linked to a proto-fascism that is predicated on family values, hard (capitalist) work, tradition, misogyny and ethno-supremacy, are functionally similar to the totemic objects that mental-health patients often cling to in psychiatric wards (for me it was a big red notebook). For those aware of the immensity of climate chaos, it is tempting to turn away from 'reality' – or to construct one of our own.

The direct effects of climate change are already having a material impact on the bodies and minds of millions. People are losing what little security they might have: access to food and water, a community, belonging to a certain place, a livelihood, family, house, home, nation, identity, personhood. When people's lives fall apart like this, the porous boundaries between self and environment perforate. The individual mind and its ecological situation start to meld. The precarity and disarray in one lead to precarity and disarray in the other. Any human thrown into an extreme situation is at a far higher risk of developing mental-health issues. As Harpreet Kaur Paul, a human-rights lawyer and climate activist, put it to me:

We can't underestimate the mental-health impacts of not knowing if family members will be found following storms, flooding, river erosion and more, or from the howling winds that shatter homes – particularly those lovingly made in informal shelters by those least responsible for climate change.

For many, this can mean long-term emotional and psychological damage, even if their material situation is later improved. Others do not make it through, not because of drowning, heat stroke or starvation, but because the mental toll is too great and suicide presents itself as the only solution.

Climate change will disproportionately impact the Global South more than the Global North, psychologically as well as physically. Sometimes this is referred to as eco-anxiety, but this is where the term can completely fall apart as a useful category. It is more helpful to think in terms of two major categories of climate-related mental-health issues.

The first, climate despair, contains most of what we have been discussing so far. This comes from relating to the problem from a relative distance, in a more abstract way. It's still potent. This is what most people mean when they talk about eco-anxiety.

The second, climate trauma, contains mental-health issues exacerbated or triggered by directly experiencing climate impacts. This is a huge deal. It may not be as trendy or as cool as the

modern conception of eco-anxiety, but this is the sharp end of the climate and mental-health dynamic. As with all good definitions, the borders of these categories are a bit blurry. Tuvalu is a prime example of the overlap.

The poster child for climate change

Tuvalu is an island nation often described somewhat offensively as the 'poster child' for climate change, given its vulnerability to sea-level rise. It is one of the smallest countries in the world, home to around 11,000 people. Most live in villages of a few hundred inhabitants. The land is rarely more than three metres above sea level. Between the wet and dry seasons, the country is thrown between tropical cyclones and droughts of increasing intensity. Some believe the islands have been inhabited for as long as 8,000 years. Tuvalu is one of the least visited countries in the world, with one flight arriving every three days.

The effects of climate change and their psychological ramifications are already apparent – climate trauma. Many of the islanders' crop yields have been seriously undermined by the increased salinity of the soil as the seas rise around them. 'Tuvalu is sinking' is a common phrase on the islands. The country's foreign minister has even begun exploring legal avenues for maintaining national sovereignty, should all Tuvaluans have to migrate and the atolls be submerged beneath the seas. Can a country still exist if it is entirely under water?

Such are the questions on the minds of Tuvaluans. Recently, a hundred islanders spoke to a researcher about their mental health in relation to unravelling climate impacts. The majority said that even abstract climate stressors – climate despair – are detrimental to their state of mind. Over 60 per cent reported at least one extreme indicator of distress resulting from thinking about climate change – *fanoanoa* (sadness), *manavase* (anxiety/ worry), *kaitaua* (anger) or 'poor health' more generally.

The mere *expectation* of future events is making these conditions worse. As one Tuvaluan said: 'Sometimes I want to sleep, but I can't because those thoughts about climate change keep popping up.' Another person said they 'hardly go out, because of those feelings'.[2] A twenty-year-old woman outlined:

> I feel really scared hearing all this on the radio about climate change. You never know if it comes true. Maybe we won't be prepared. We just sit there and wait to die.[3]

Climate change is going to threaten the psychological resilience of everyone on the planet. It can hit us in the abstract, from a distance, and it can hit us with concrete climate shocks in our communities. Through 'eco-anxiety', or by overhauling material circumstances, it can seed depression, generalised anxiety, PTSD, psychosis, prolonged grief, dissociation, crippling phobias, eating disorders, schizophrenia, bipolar disorder and suicide. It is already happening.[4] The tendrils spread menacingly. Climate change will increase poverty, which will impact people's mental health. It will increase inequality, within and between countries. It will create food shortages, which drove people mad over the course of the pandemic and has done throughout history. Heatwaves trigger domestic abuse, alcoholism and gun violence. Flooding decimates and depresses. Conflict and natural disasters could mean over a billion climate migrants by 2050. Migrants are some of the most likely people to develop mental-health issues. They are also the least likely to receive support. Less than 10 per cent of nations have any explicit plan to address mental-health-related climate impacts.[5] That is, if you will excuse the looseness of my language, completely insane.

Ignoring mental-health problems caused by climate impacts is dangerous. We can't continue to magic them away with cognitive sleights of hand that deem the lived psychological experiences of those most impacted to be 'natural.' Studies suggest that the psychological impacts of natural disasters can outweigh the physical by a factor of 40 to 1. Meanwhile, the provision of mental-health services in low- and middle-income countries is scarce, to say the least. Most of the meagre resources are spent on in-patient psychiatric hospitals or asylums.[6] Around a third of countries have no state-funded explicit mental-health provision at all.[7] Asylum seekers in some Global North countries are five times more likely than residents to experience mental-health issues. A massive 61 per cent of asylum seekers in the UK report extreme mental distress. They are, at the same time, far less likely than British citizens to receive any support.[8] This all folds into a neocolonial narrative of dehumanising people of colour that goes back centuries.

Mental health is often seen as a Global North issue, as though the state of our minds is only consequential once a society has reached a certain level of economic development. It is as if it is a privilege to be able to care about the psyche. Even in rich countries, a lot of people see therapy as indulgent and egotistical. Psychoanalysis is narcissism. People who claim to have mental-health problems are seen as weak, overdramatic or attention-seeking. Thankfully, there has recently been something of a destigmatising and demystifying of mental health, but there is still a stubborn sense that it is what we should focus on last – the cherry on the cake. We should only explicitly confront mental-health issues if they are making life unliveable – and people affected that badly are usually othered, put into a separate category of human of which we should be wary. As a result, mental-health problems are deprioritised. One way that this happens is by relegating mental-health concerns such as major depressive disorder, anxiety and PTSD to normal, expected behaviour that is the natural result of trauma, and therefore nothing to worry about. In the Global South, challenging mental-health issues are often assumed to be normal responses to suffering, and therefore somehow 'natural,' 'inevitable' and 'less significant'. There is more than a whiff of racialised dehumanisation at work here.

These ideas are from the same worldview as the horrible notion of social evolution, so loved by the Social Darwinists. This is more commonly referred to as 'stages of history', with humanity progressing teleologically from a 'state of nature' to 'civilisation'. Karl Marx delineated these stages as Prehistory, Classical, Middle Ages, Early Modern and Modern. Adam Smith did something similar, calling them the Age of Hunters, the Age of Shepherds, the Age of Agriculture and, finally, the Age of Commerce. These deterministic stages of social evolution suggest humanity's development from 'wild' to 'modern'. The latter is, by conventional definition, more developed than the former. As such, there is a disregard for perceptions of the universe in any culture other than those classed as modern or civilised (or, in other words, 'industrial capitalist' or 'cognitively white'). These are derogatory, dehumanising and ultimately exploitative ways of framing the world.

Maslow's hierarchy of needs, from 1943, is a similarly constructed touchstone. Maslow, it should be remembered, was the academic who refused to study those with mental-health problems because the study of crippled, stunted, immature and unhealthy specimens could yield only 'a cripple psychology and a cripple philosophy'. His pyramid starts with 'physical needs', building up to 'safety needs', then 'love and belonging', followed by 'esteem' and finally 'self-actualisation'. Those categories at the top of the pyramid are deemed the highest states of being, the fields of most advanced psychological wellness, as well as being fundamentally dependent on the establishment of the categories below them. I can find value in the idea that we should prioritise physiological needs and safety, but there is an unpleasant suggestion here that one's psychological needs are not even worth considering until other basic needs are met. The pyramid is in fact explicitly labelled in this way.

Just as social evolution's stages of history look down upon previous modes of existence, Maslow and those still influenced by his work seem to be working on the assumption that the human mind does not really matter until a person has reached a certain level of material stability. But we know, irrefutably, that everyone feels. To suggest that the emotional and mental existence of the materially deprived is not worth paying attention to has the effect of stripping an individual of their humanity. In Pakistan, if somebody has lost their home because of flooding that has covered a third of their nation, their psychological wellbeing is likely to be in tatters (see Chapter 4). These situations are the ones in which we are most likely to see mental-health issues. People suffering in this way deserve dignity and solidarity. They certainly require material support. The psychological and the material, however, cannot be made synonymous. Mental health is not a 'Western luxury'.

In their extraordinary book *Inflamed*, Rupa Marya and Raj Patel write:

> Truly holistic health must contend with the elements that continue to make all people unwell. Locating the disease-causing entities in the social structures and the grave misunderstandings that created them. Systems that position humans as supreme over the entire web of life, settler over indigenous, a singular religion over all other world views, male over female and non-binary understandings of gender, white over every other shade of skin – these must be dismantled and composted. We must reimagine our wellness collectively, not simply as individuals or communities, but in relation to all the entities that support the possibility of healthy lives.[9]

Around 60 per cent of India's population works in agriculture. Despite this, the industry is responsible for less than a fifth of national gross domestic product (GDP). Around 85 per cent of India's farmers work on smallholdings, many on the brink of subsistence. This is a multi-causal problem, of which climate change is a key feature. The government gives a small amount of money directly to farmers but that rarely covers water and electricity. Moneylenders in rural India charge extortionately high levels of interest. There are few alternatives and debt levels are high. For farmers just about staying afloat, crop failure can be catastrophic.

A landmark paper published a few years ago found that, in the thirty years leading up to 2013, 60,000 farmers in India died by suicide. These were just the suicides explicitly linked to climate change, low rainfall and crop failure having pushed people over the edge of despair. The paper's author, Tamma A. Carleton,

Assistant Professor at the University of California Santa Barbara, believes that the suicide death toll is probably much higher.

Since the publication of Carleton's widely disseminated findings, things have only worsened. India's populist prime minister, Narendra Modi, initiated reforms to open agriculture up to the free market. He had promised to double farmers' incomes, but the reforms were set to strip farmers of their already meagre protections, such as minimum price guarantees for key crops. In late 2020, hundreds of thousands started protesting. They set themselves up outside Delhi, remaining a forceful presence for over a year. In November 2021, Modi bowed to the pressure and repealed the laws.

From 2019 to 2021 alone, reported suicides by farmers increased by 30 per cent. Ketki Singh, Vice President of the Bhartiya Kisan Union's women's wing told *Deutsche Welle* that: 'Climate change has acted as the last nail in the coffin . . . Can you imagine that nearly 30 people in the farming sector die by suicide daily?'[10] Indian farmers' incomes have remained stagnant for five years. Prices, on the other hand, have rocketed. In 2021, 5 million hectares of crops were lost due to climate impacts.[11] By 2030, crop failure is predicted to be 4.5 times more likely. By 2050, it will be 25 times higher.[12]

Fanning the flames of conflict

Food insecurity is one of the principal ways in which the Intergovernmental Panel on Climate Change (IPCC) has linked climate change to war, alongside water scarcity and migration because of natural disasters. Some scholars, media outlets and public figures, including King Charles, argued that the ongoing conflict in Syria was the first 'climate war'.[13] Extreme drought in the region, the argument goes, led to large-scale migration that has in turn precipitated conflict. Academics have since challenged this view, or at least questioned the strength of the link.[14] In 2015, an influential paper made the bold claim that for every one degree of temperature rise there will be a rise of 11.3 per cent in intergroup conflict.[15] But it's not that simple. As one climate and conflict scholar tells me: 'this rhetoric around climate wars is misplaced.' They said that in West Africa, for instance, climate change is a factor that squeezes the opportunities of those already in precarious situations. Food insecurity can lead to generalised insecurity, which boosts recruitment into armed groups. So does land and water scarcity, alongside financial insecurity. Global capitalism, according to those studying the area, also drives home a desire for the status attached to material wealth. This makes the money offered by militias even more enticing. Climate change can tip people over that threshold.

Government action to address climate change can also, confusingly, precipitate conflict. In Mali, the state has shrunk in response to demands from Western institutions like the International Monetary Fund (IMF) and World Bank, but the global discourse around deforestation and reforestation means that the forestry service has maintained its budget. As a comparatively strong department, it has also become more militarised. They stop citizens from accessing forests as they would a commons, attacking those trying to harvest trees or graze animals. The forests have historically been a place of refuge, as well as a site of nourishment in times of crisis. Now, the state is denying access due to Western-imposed environmental edicts. I am told that jihadists are using the actions of the forest service as a recruitment narrative – and that they are doing it successfully.

On average, one in five people living in conflict areas experience some form of mental-health issue. More than 5 per cent suffer from bipolar disorder, schizophrenia, severe depression, severe anxiety or severe PTSD.[16] That is more than double the level of the average population. PTSD is often associated with conflict, but the focus has mainly been on combatants rather than on civilians.[17] There are now over 70 million children worldwide living in conflict zones where we know that militants have perpetrated sexual violence against them. The risk of sexual assault and abuse against children in conflict zones is ten times higher than in 1990. In Yemen alone, 83 per cent of children, over 11 million, are at risk of sexual violence.[18] Unsurprisingly, child victims of sexual abuse are around three-and-a-half times more likely to develop serious mental-health issues.[19]

Physical violence aimed at children, whether by combatants, their parents or strangers can also have serious impacts later in life. Poverty, food shortages and the stress of constant itinerant movement all lead to higher rates of physical violence. Under such circumstances, people can snap and beat children. They are so at risk in conflict zones that academics and clinicians often describe their experiences as 'polyvictimisation'.[20] Children who experience neglect and abuse can be four times more likely to develop bipolar disorder, schizophrenia and psychosis.[21]

Heatwaves alone can also trigger higher rates of abuse, trauma and mental-health issues. In Madrid, one study found that higher temperatures led to more intimate partner violence. The most extreme instance found a city-wide hike in domestic abuse and femicide of 40 per cent.[22] Worldwide, domestic abuse is the leading cause of morbidity and death amongst women of child-bearing age. Astonishingly, most of the recorded burden comes from the psychological effects of abuse, rather than its physical effects. This usually comes in the form of depression, insomnia, PTSD, suicidal thoughts and suicide.[23]

Meta-analyses of social-media posts during heatwaves show a reliable increase in depressive language across the board.[24] Hotter temperatures lead to more episodes and diagnoses of schizophrenia, mood disorders and neurotic disorders.[25] People with bipolar disorder, a profoundly climate-dependent condition, are more likely to have so-called relapses when the mercury shoots up.[26] Combined with other stressors this can be brutal, especially in already hot countries. Religious practices, such as fasting during Ramadan, can escalate the risk of someone with bipolar having an episode by almost three times.[27] In the US and Mexico, just the temperature increase caused by climate change could mean up to 40,000 additional deaths by suicide by 2050. This is comparable, the report says, to the estimated impact of economic recessions or of the absence of suicide-prevention programmes or gun-restriction laws.[28]

Higher temperatures lead to more public violence, too. Riots are more common.[29] So are murders. One study predicts a 6-per-cent rise in homicides globally for a rise in temperature of 1 degree Celsius.[30] A study in Chicago found that a temperature increase of 10 degrees Celsius above the average led to a 33.8-per-cent rise in shootings per day.[31] Police are also more aggressive, tense and quicker to make negative assumptions about suspects when it's hotter than normal.[32] One experiment found that in a controlled environment, just raising the ambient temperature of a room from 21 degrees Celsius to 27 degrees Celsius increased the police's tendency to fire their weapons by 65 per cent.[33] In a potentially dangerous mix, natural disasters can also dramatically increase police alcohol consumption. Following Hurricane Katrina, police drinking went up from an average of two drinks per day to *seven a day*.[34] Victims of gun violence, whether directly in the line of fire or not, are far more susceptible to depression, intrusive thoughts, personality changes, anxiety, insomnia and PTSD.[35] The burden of this will be disproportionately felt in communities that are already the most vulnerable to rising temperatures.[36]

Wildfires hit children and young people in particular when it comes to PTSD, nightmares, depression and insomnia. Burning forests, houses and thick smoke trigger so much fear and stress that the body has an extreme reaction.[37] Survivors of Australian bushfires, particularly those closest to the flames, are likely to suffer psychological effects that can last for years.[38] Some research even suggests that breathing in wildfire smoke can itself drive an increase in depression, irritability, insomnia, fear, hopelessness and lethargy.[39] Whilst recent Californian wildfires have rightly been extensively covered in the news, Indonesia has been experiencing conflagrations just as large for a decade, also visible from space. The fires there now recur annually. They usually result in smoke so widespread that the cloud casts a hazy pallor over most of the nation, even spreading to neighbouring Singapore and Malaysia, setting record-high levels of air pollution over state lines.[40] In southern Sumatra and Borneo, the Pollutant Standards Index has even hit a terrifying 2,000,[41] where a reading of 350 or above is deemed hazardous to health. There are copious amounts of research linking poor air quality to cognitive decline, Alzheimer's, dementia, depression and suicide.[42] In 2019, the UN warned that there were 10 million Indonesian children at risk of severe air pollution.[43] On its own, being forced to live in a hazy orange cloud of suspended burnt matter is psychologically brutal. Recent images of incinerated trucks, soot-blackened children and red skies from across the world are all unsettlingly close to real-world manifestations of end-of-days hellfire and brimstone.[44]

In some areas, such as in recent US fires, around 10 per cent of those in the vicinity stay behind to protect their homes and fight the flames. These citizen firefighters are at high risk of mental distress, along with inmate firefighters like those in the state of California, prisoners who are deployed as unpaid emergency labour. They are promised a two-day sentence reduction for every day they put themselves in front of infernos. Between 10 and 20 per cent of conventional wildland firefighters develop PTSD.[45] It is like throwing people into a war.

Even in the Global North, flooding that is minor in comparison to the threatened sinking of entire nations is already having noteworthy effects on people's minds. A study published by the UK government found that victims of flooding are six times more likely to experience mental-health issues, including PTSD, depression and anxiety. These psychological impacts can last for years.[46] One reason for this is a loss of a sense of community, an important part of feeling that we belong, even in a nation as socially stratified as Britain.[47] One in six homes in England is currently at risk of flooding. In 2021, the UK saw some of the most severe flooding in living memory. In London alone, the fire brigade received more than 1,000 calls related to flooding. Cars were plunged under water and homes evacuated. The same was true across much of Europe, with some places seeing the heaviest rainfall in a thousand years. As we shall see in succeeding chapters, the psychic toll of catastrophic flooding in countries like Nigeria and Pakistan is incomparably huge. The IPCC predicts we will see at least 1 metre of sea-level rise by 2100.

Indigenous insights and injuries

Solastalgia is the sense of loss, woe and shift in identity resulting from seeing one's environment change in front of you. It is a term now regularly applied to some indigenous peoples' experience of environmental change. For centuries, the Inuit have had to weather externally imposed anthropogenic changes to their cultures and landscapes. The Canadian Arctic, where some Inuit communities have lived continuously for 5,000 years, is one of the fastest-warming places on earth, described by locals as the weather behaving *uggianaqtuq* (strangely and unpredictably). The weather is unfamiliar; perhaps, even, no longer family. The psychological toll this takes is huge. There has been a marked increase in depression and addiction. As one Inuit community member shared with psychologists, changing snow, ice and weather: 'can be hard on your mental capabilities. It can be hard to keep your sanity.'[48] If sea ice is a key source of autonomy, as well as mental, emotional, spiritual, social and cultural health, then the loss of ice can be experienced as mourning, as well as a traumatic severing of links with one's ancestors.[49]

There are worryingly few cultures, in terms of historical diversity and percentage of world population, that still engender a knitted, stitched-in way of equitably interrelating with non-humans. Globally, indigenous people account for around 6 per cent of the population, but together they steward 80 per cent of the world's biodiversity on just 20 per cent of the world's land. Whilst it is important to note that indigenous groups, internationally, are by definition hyper-diverse (they account for the majority of humanity's cultural diversity in and of themselves), on average they are far more at risk from the environmental impacts of climate change and already suffer disproportionately from mental-health issues. This is perhaps unsurprising, given the concerted and ongoing colonial and neocolonial missions to eradicate the basis of their subsistence.

Virtually everyone else, ecologically, is left alone. They spend their lives staring uniformly at the inventions and products of human civilisation, a uni-species echo chamber, a hall of mirrors resulting in an infinitely reflected monocultural hegemony. This is both navel-gazing in the extreme, and a deeply companionless way to live, deaf to the cacophony of voices speaking from the rest of the non-human living world. As Robin Wall Kimmerer, scientist, author and member of the Citizen Potawatomi Nation, writes in her book *Braiding Sweetgrass*:

> Philosophers call this state of isolation and disconnection 'species loneliness' – a deep, unnamed sadness stemming from estrangement from the rest of Creation, from the loss of relationship. As our human dominance of the world has grown, we have become more isolated, more lonely, when we can no longer call out to our neighbors.

Climate change is a further rending of this fabric. Human and non-human lives, as well as the relationships between the two, are in the firing line. Climate change was born of disconnection and domination, but is perhaps the final cementing of Western modernity's supposed primacy over nature, simultaneously ripping people from the means to survive independent of extractivist culture. It could, on the other hand, be the forest fire that allows new shoots to grow. The impacts of climate change are further decimating the cultures least responsible for the crisis, but they are the ones best placed to school the rest of us. A warped perception of the universe is, then, threatening to further warp the minds of every human on earth through the vector of our climate.

For indigenous communities – whether across the melting icescapes of the Inuit, the burning and ransacked ecosystems of the Amazon Basin, the scorching heart of Aboriginal Australia, the thawing permafrosts of the Sa'ami, Komi and Nenet, the polluted and deluged Niger basin of the Ogoni – land and water are often a vital part not just of identity and interrelatedness, but of genealogy. It is no surprise that indigenous people comprise the bulk of environmental defence movements. They are also more likely to be attacked, imprisoned and killed by the authorities for standing up to the interests of economic power.

Unsurprisingly, the vast majority of studies looking at the mental-health implications of climate change focus on events and populations in the Global North. The IPCC's sixth annual report was the first of its kind to explicitly dedicate space to mental health. North America had pages dedicated to it. The African chapter had just one sentence. Dr Elaine Flores is one of a team of academics explicitly dedicated to studying mental health and climate change in the Global South. They have reviewed thousands of papers, and only a sliver of them, she tells me, are sufficiently detailed to reliably link climate change, resilience and mental health. 'We need more research,' she tells me, 'but there are strategies that are working.' One of the key findings of Dr Flores' work is that communities with a higher level of resilience, a more embedded sense of community and togetherness, are generally better at responding to climate shocks. Mutual aid plays an important role, as do other informal support networks.

In 2004, Sri Lanka was hit by a devastating tsunami. Hundreds of Western psychologists flew to the country to provide mental-health support. To their surprise, they found a profound lack of worrying mental-health issues. Many chalked this up to denial. In his book, *Crazy Like Us: The Globalisation of the American Psyche,* Ethan Watters argues that many of our beliefs about mental health are self-fulfilling prophecies. In Sri Lanka, according to those Watters spoke to, the strong sense of community, experiences of economic hardship, history of colonialism and the religious infra-structure of Hinduism and Buddhism meant that the people were readily able to 'integrate and give meaning to terrible events'.[50] The Centre for Suicide Prevention agrees, saying the psychological resilience of the Sri Lankan populace post-2004 was largely the result of 'a very strong sense of social connectedness'.[51]

The situation after Typhoon Hainan, which hit the Philippines in 2013 and killed 6,000 people, was quite similar – at least from the perspective of Western medics. The prevalence of PTSD did not appear to rise following the disaster. There was, however, a considerable increase in 'nonspecific' psychological distress.[52]

Doctors did not know how to categorise this. Along with the insights of many living in the Global South, this points to something important. Not only is mental health in the Global South not prioritised by the international community, it is also only understood in Western terms. These terms are not necessarily relevant to, or appropriate for, those with different conceptual frameworks of the world, society and mind.

Airdropping Western biomedical models of psychiatric diagnostics and support into the Global South have limited use and are potentially dangerous, if not exploitative. 'Just providing more clinics and psychiatrists won't solve the problem,' Dr Flores tells me. 'We've got to start contrasting the atomised mental-health support on offer with social models of illness and care.'

Mental-health issues that come from climate chaos are indictments of the system. But we don't necessarily need any more indictments of the system, what we need is tools. The different mental states we are thrown into can, paradoxically, catalyse the discovery of different worlds, as can our related survival strategies. The socio-economic system we live in leaves our imaginations tightly constrained. It amputates a diverse series of future pathways. Living as we do involves folding ourselves into the dominant culture's pre-defined narrative of existence. This results in a trap. On an individual level, we are taunted by the claim that we can be anything. On a societal level, we have the opposite: they tell us we have no choice but to live this way, no matter how perplexing, perverse and unjust it may seem. But already, climate change means that none of us are in Kansas anymore. We are somewhere else new and scary. But madness and our responses to it can force an inescapable awareness of the sheer number of alternative futures we could inhabit. Insane as it may seem, like bubbles expanding against the meniscus of our conscious selves, the worlds we want may not be as far away as we think.

PART TWO

4

Nigeria: activism and renewal

'I tell you this, I may be dead but my ideas will live on.'

Ken Saro-Wiwa

IN 2022 NIGERIA WAS reeling from devastating flooding. The inundation, caused by climate-induced heavy rainfall combined with a mistimed release of water from Lagdo Dam in neighbouring Cameroon, hit thirty-four of Nigeria's thirty-six states. More than 3 million people were affected. One and a half million were displaced.[1] Vast swathes of farmland were destroyed. Three-hundred thousand homes eviscerated. More than 600 people died. The floods exacerbated an already precarious food crisis, and many communities were accessible only by boat many months later, even as the waters were receding.

The WHO tells us that people forced from their homes are significantly more likely to suffer from mental-health issues.[2] This is unsurprising. The scale of the impact, however, is huge. This is not much studied, and data on mental health in the Global South is rarely even documented. One rare piece of research, on the opposite side of the continent to Nigeria, found that more than 50 per cent of internal migrants in Sudan suffered from major depressive disorder, generalised anxiety disorder, social phobia or PTSD.[3] Lack of treatment can easily ensure these conditions become chronic. Mental-health first aid is slowly starting to be seen as a useful response to crises like these, but there remains a sense of it as something of a luxury that is subordinate to material aid. Thousands upon thousands are left wrestling with despondency, despair and derangement.

The last few years have not been kind to Nigeria. It fared comparatively well on the health front during the pandemic. Ebola, as the South African comedian Trevor Noah jokes, prepared the continent for a ship-shape pandemic response long ago. But the early stages left Nigeria with its worst recession in four decades. Now, the country's inflation rate is above 20 per cent, the highest level in almost twenty years. Food and fuel prices are soaring. Inequality and poverty are rising, with the poorest being hit hardest by the increasing cost of staples, depreciating wages and lack of access to savings and assets. The country stands on the brink of its third recession in seven years. Flooding means millions of people, mostly the rural poor, have lost their homes, their livelihoods, their security – and all of this in the midst of economic chaos driven principally by the vagaries of international capital.

Climate change is rarely mentioned as a cause by the people most affected by the rising waters. Most of those living in rural settings have no access to the internet, which is the primary means of accessing information on climate change in the country. Rural Nigerians are more likely to reach to the divine for explanations of natural disasters. From their position, climate change may not be relevant or useful as a framework for explaining the world. What matters is confronting the consequences.

Urban Nigerians are more aware as climate change is in the news a lot, is an important factor in elections and the President often talks about the issue. There is a huge amount of support for international loss and damage payments from the Global North, alongside demands for climate adaptation and increased resilience. Oil also comes into the equation. Nigeria is in the top ten countries in the world for proven oil reserves, and *half* of the Nigerian Federal Government's revenue comes from oil.[4] Much of this is sucked out of the ground beneath the Niger Delta. Decades of extraction and repeated spills have made the Delta one of the most polluted places on the planet.[5] One of the biggest operators in the region has been Shell, a transnational oil and gas company based in the UK, the same country that colonised Nigeria and ruled over it for a century. This one company, on its own, has released over 17.5 million litres of oil into the Delta's ecosystems.

There is a strong and growing climate movement in Nigeria. Young people have clearly been at the forefront, as with much of the contemporary climate movement worldwide. Nigeria is a young country in both senses. It gained independence from the British in 1960, but in addition, 70 per cent of Nigerians are under the age of thirty. The young are driving the climate movement, as well as a population boom.[6] Already, Nigeria is both Africa's largest economy and its most populous country. Many claim that this puts the nation in a uniquely powerful position to help define how Africa, and the Global South more widely, responds to this era of polycrisis. Young people are pioneers who must learn to survive and thrive in an unpredictable, kaleido-scopic, ever-changing environment. Many of the lessons of the past have become anachronistic – we are in alien territory. It is easy, and understandable, to isolate in the face of disaster but some find ways to face it. Jennifer Uchendu is a prime example.

Jennifer has been working on climate change for years. In 2016, she founded the organisation SustyVibes in Lagos, a youth-led movement to educate people on climate change, organise nature clean-ups and tree planting. They also conduct research on climate solutions and, vitally, make sure that the fight for climate justice is as enjoyable as possible. SustyVibes, Jennifer tells me, is 'even in the name, a lot of fun. Let's talk about sustainability in fun ways. Let's use pop culture to get everyone on board. Let's not make this, you know, gloomy and sad.'

Alongside its other activities, SustyVibes hosts parties. These have educational workshops and talks alongside art exhibitions, dance groups, open-mike poetry plus food and drink for everyone. 'The parties have been a way to change how the conversation usually goes sideways,' Jennifer tells me. 'We want everyone to come to the party excited that they're young people making change, and to leave feeling excited, connected and inspired. That's what every Susty party aims to do.'

Most funding organisations do not see a party as a valuable use of resources. They would rather fund a seminar or lecture series, or donate equipment. But Jennifer knows from experience how fundamental it is to celebrate together and to look after each other. She also knows what can happen when you don't do that.

In 2018 Jennifer ended up having lots of conversations with her friends about mental health. 'We already had a sense back then that there was a psychological impact of the work that we did. On the one hand, every time we came together as a community of volunteers and members, we were able to draw strength and build to the next action.'

But there was also a more worrying, sombre side. 'We just felt a sense of powerlessness, like we were playing a losing game whilst you see your peers move up the ladder in their careers, buying cars, moving on, seemingly unaware. And you're just stuck as a climate activist, continuing to tell people to do better, trying to hold governments to account. It felt like a drop in the ocean.'

The team were also starting to talk about how climate was affecting their minds more systemically. There was a deep despondency, despite their best efforts. But at that stage, Jennifer tells me, 'it was very vague. It was just a suggestion. People were asking if it was possible.'

SustyVibes decided to host an event for young people to explicitly put the question out to the community. Was there a connection between climate change and mental health? They were hit by an unexpected outpouring of emotion. 'We didn't have a name for what we were experiencing yet,' Jennifer says. 'But that gathering gave us a form of validation. It was a space for more and more people to say "yes, I feel this too". We could experience the emotions collectively. That was really important.'[7]

UN conference disillusionment
The following year Jennifer went to the UN climate talks in Madrid. It was her first Conference of the Parties (COP). It was devastating.

She saw first-hand that the leaders had 'lost their passion and didn't really care'. She and I talked about this for a whilst. I remember my own sense of shock when I was at my first COP in Copenhagen, getting to the conference and seeing an almost aggressive level of inertia and apathy. 'It was very similar for me,' she said. She told me that people tend to think of the UNFCCC as the 'final level' of climate activism, like reaching the final boss in a videogame, or pulling back the curtain to reveal the wizened old Wizard of Oz. I remember that reveal, too. It felt a bit like the gut-punch a child receives when they work out that their parents are fallible. In truth, no one really knows what they're doing. But at COP, it is even scarier. These are the people reportedly in charge of averting climate chaos. When you step into the hall and see their blasé approach, well, it can be nightmare fuel.

'Madrid,' Jennifer says, 'was peak eco-anxiety.' There was also a lot of tokenism and greenwashing at the conference. Virtue signalling, rather than any genuine commitment to change.

> I've always heard about it, but I've never really seen it. I'm an activist, so I just go into field, do what I need to do and that's it. It was confusing for me. I thought I was just going to see a lot of action. A lot of climate activists, particularly in Africa, aspire to go to COP, to be a top negotiator, globally recognised. It's almost like gaining celebrity status. I was funded to be at COP that year. But I just thought: why? I could have used that ticket money to run a project at home.

That, she tells me, was 'very, very heavy to bear'. She was also painfully aware that many of the systemic drivers of climate change still play out in the process itself. 'Greed,' she says, 'is still perpetuated in events like these. Everyone is really looking out for their own agenda.' Extractivism, whiteness, neoliberalism, neocolonialism, racism, sexism, ableism – all are in evidence.

After the climate talks Jennifer felt numb and powerless. 'I was devastated. I was crying every other night.' She questioned, over and over, whether it was worth continuing the work to which she had dedicated so much of her life. 'Why wallow in this anxiety?' she wondered, when instead she could find a job with short-term but meaningful gains. 'Then at least I could overcome the poverty spell that so many young people in this part of the world are bound to.'

But Jennifer realised that the very fact that she was so moved, the sheer power of her emotional response, could itself be proof of the importance of her work. You can't be heartbroken unless you love something, after all. Jennifer managed to turn her pain into focus. She started investigating what exactly was affecting her. 'Why do I feel so deeply?' she asked. 'Why am I so concerned about this issue in particular?'

She trained her sights on climate emotions for a Master's dissertation. She decided to look at the psychological dimensions of the climate crisis in different contexts and demographics across the UK, the country in which she was studying. She looked at families, classrooms, interactions around the dinner table. She also focused quite heavily on Extinction Rebellion (XR). The older members of XR, Jennifer found, were very resistant to the movement being anything but predominantly white. They seemed opposed to incorporating different voices. 'They were unable to see that it's the very ideas of whiteness and power that brought us to where we are,' Jennifer says. The young, by contrast, wanted a more inclusive movement, one that saw climate justice and social justice as two sides of the same coin. This conflict between young and old, Jennifer found, was a source of a great deal of psychological difficulty for the young. There was guilt and shame piled on top of the guilt and shame that was already there.

At first, Jennifer was shocked that her Western counterparts might also be experiencing anything like eco-anxiety. Initially, it wasn't something she could relate to. 'Life's a bit easier in the UK than in Nigeria. In the UK there's something of a safety net, and compared to us, the UK will be relatively fine in the climate crisis. I was filled with a lot of mixed emotions when I found that even there people were experiencing mental-health issues because of the climate.'

Jennifer's work ended up distinguishing her own experience at COP and those of her contemporaries in Nigeria, from what she found in the UK. Those in the West, she wrote, are often suffering psychologically from a set of emotions based on guilt and shame, whilst the mental-health ramifications of climate change in countries like Nigeria are more frequently situated in rage and anger. 'But we should remember that they're just entry points,' she says. 'People experience lots of different emotions when it comes to the climate crisis. There is the underlying problem, the pathway, then there are acute situations like depression, PTSD, or suicidality. They can end up being chronic. What we need to realise is that it's something systemic.'

After her Master's Jennifer returned to Nigeria and founded TEAP: The Eco-anxiety in Africa Project. Despite the fact that SustyVibes had grown to nearly 500 active volunteers, she found it difficult to find much financial backing for TEAP from outside funders. 'There's already little support for climate work in Nigeria, and even less for mental health. When you combine the two topics it's a completely different ball game.' Jennifer wanted to dig deeper into what the connections were between planet and mind, as well as focusing on academic research. TEAP began by hosting in-person listening cafés, monthly webinars on climate and mental health for young people across the continent, and creating as many opportunities as possible for people to come together and share their stories. 'Climate seems almost like a ticking time bomb, but people are already unravelling, psychologically. One of the most important things we can do is approach young people and let them know that their concern, their distress is valid – because the world is literally falling apart.'

Young people in Nigeria are already in a delicate position in mental-health terms. Around a million people leave the country every year. Meanwhile, there are millions of internally displaced people, largely because of climate events, conflict, or a combination of the two. The Islamic terrorist group Boko Haram have been active in the north-east region of the country for more than a decade and over 5 million people have been displaced by the insurgency. Although the relationship is complicated, climate has played a role in exacerbating the conflict by increasing communities' economic insecurity and incrementally stripping away their resilience. At the same time, the Nigerian government's own figures put the level of multi-dimensional poverty at 63 per cent for the whole population. More than half of the country's poor are under the age of seventeen.[8] Nigeria has some of the lowest social spending on the continent, despite there being significant untapped opportunities for the government to collect more in taxes. According to Oxfam, the combined wealth of the five richest Nigerian men could *end* extreme poverty at the national level.[9] Nigeria regularly comes bottom of the international league table measuring countries' attempts to address inequality.[10]

Given all of this, it might seem like a distraction to start talking about the mental-health implications of climate change. But more than 50 per cent of Nigerians aged 16 to 25 are either 'very worried' or 'extremely worried' about the climate. That comes on top of everything else, and compounds the angst which is often a trigger for things that are much worse.

Climate change [says Jennifer Uchendu] is this social planetary health problem, happening literally everywhere on earth. Everyone has a feeling about it and a relationship to it. But we have to remember it's also about the power structures that caused it, which affects how we internalise climate change and how we act. These dynamics are playing out within families, often through anger at older generations. There's a sense of betrayal, too. Betrayal by government. Betrayal by business. Young people often experience intense climate and mental-health effects because they're morally injured. At the same time, they feel like they don't have any real agency.

When I asked if climate change felt like a distinct issue in terms of mental health, she gave me a 'yes, and no'. 'There are different layers of anxiety,' she said:

> Some of them are connected to climate, like food insecurity or sometimes conflict. But in Africa, I'd say we're kind of used to disasters or, let's call them 'problems', problems you might not experience so much in the West. But what makes climate change distinct is that at the same time as it's making all of that worse, it's driving up a lot of the most acute problems and creating new ones. There's an upheaval, even a loss of identity. If you're displaced by the climate crisis it doesn't just feel like a bad thing you can navigate, it affects your entire destiny and future. Young people, partly through increased internet access and social media, are seeing that and expecting it. So, we create spaces where it's okay, normal, to be vulnerable. What we're doing together is finding ways to transform the fear, the depression, the anxiety and anger into hope. Hope and action.

One of the many organisations TEAP works with is a youth-led mental-health NGO called the Mentally Aware Nigeria Initiative (Mani). TEAP and Mani have held workshops together, making sure that there are mental-health professionals available to facilitate and guide sessions, as well as helping connect people with additional support, should they need it. What can often happen in the climate movement[11] is that, even if the topic of mental health is raised, the conversation that follows can leave people in the room feeling exposed and vulnerable – and that response often does not meet with sufficient compassion, action and support.

This is not a snowflake issue, it is about making sure people feel comfortable peeling back layers of distress and trauma, which only happens healthily if they can trust that what is shared will be respected. Knowledge and experience are needed to avoid participants overexposing themselves. I have personally done this, accidentally. I have also seen it happen to others more than a few times: individuals sharing deep, dark and personal thoughts, feelings and traumas, only to be met with a head-tilt, a tokenistic 'thank you' and then having the session move on as if nothing of consequence had been shared. Understandably, this can repel people from these spaces. Jennifer knew from the outset that safeguarding was important, having worked with young people for a long time. TEAP works with counsellors, psychologists, therapists, psychiatrists and other mental-health professionals to inform the structure of its work.

TEAP recently ran a project to curate the voices of people suffering the mental-health impacts of climate change. The aim was to put on an exhibition, but some of the entries were quite grim and required direct contact. One young man wrote in saying he was suicidal. TEAP connected him with a psychotherapist, one who was very aware of climate psychology. That was just the beginning of a process. They explicitly did not want to individualise his suffering – quite the opposite. Together, the young man, the therapist and the team at TEAP worked to connect him with a growing community of activists. They wanted to let him know he was not alone. Sometimes, Jennifer tells me: 'It's possible to redirect some of these feelings into community action. People should know there's not something particularly wrong with them, but that it's something we're all going through. The individual who wrote in saying he was suicidal, for instance, is now very active in the community. He has found an outlet through photography and is coming to lots of events.'

What kind of future do we want?
A huge dimension of mental-health issues related to climate change result from feeling isolated, from having the sense that you can see the dark future and the rest of the world is continuing, uncaringly. That breeds dissociative feelings and even threatens to drop the bell jar over you, separating you from everyone else's contrived reality. 'I've always advocated for space,' Jennifer says, 'space to support validating these experiences, space for more conversation and dialogue, space for people to just talk about

how they feel.' Part of that, according to TEAP's methodology, is to dedicate time to imagining radically better worlds. What, they ask, do we want to achieve? What kind of future do we want? How are we going to get it? A lack of agency, as we know, is often at the root of resignation. We need to flip that on its head and lay the groundwork for empowered action.

Organisations like TEAP and Mani have their work cut out for them. Nigerians are typically reluctant to share concerns about mental health in public. Two of Mani's main focuses are mental-health stigma and discrimination, still widespread across much of the continent. One of President Muhammadu Buhari's final acts before leaving office in 2023, following the recent presidential election, was to sign a new mental-health act into law after multiple failed attempts.[12] The bill has been twenty years in the making. None of the previously proposed legislation came to fruition. Technically, the legal infrastructure that Nigerians have had to rely on for mental-health treatment was, until now, a law from 1958, introduced during British colonial rule, referred to as the Lunacy Ordinance.[13] The Ordinance gives the state the right to imprison those with mental-health problems, although the text prefers to use the term 'mentally ill' (a marked improvement on 'lunatic' and 'idiot', as used in earlier iterations of the text). The Ordinance not only allowed for the 'mentally ill' to be detained, but also those found 'wandering at large', 'committing some offence', or a person believed to be 'about to commit some offence against the law'. In a similar blurring of the lines between health difficulties and criminality, the Lunacy Ordinance deems suicide illegal. If someone attempts suicide and survives, only to have someone press charges, they can then, technically, be incarcerated.

One in four Nigerians are estimated to suffer from mental-health problems, 80 per cent of whom have virtually no access to treatment.[14] The country currently has fewer than 300 clinical psychiatrists. Around 3 per cent of Nigeria's already meagre health budget is dedicated to mental health. Some 90 per cent of that is allocated to psychiatric hospitals, infamous for their stark conditions.[15] Early intervention is extremely difficult because the likelihood of treating psychiatric issues as something deserving compassion is low. Those who make it to state-run 'rehabilitation centres' and psychiatric wards are not hugely fortunate, either. Despite some pioneering attempts to humanise mental-health issues, such as the Aro model in Ogun state,[16] the Nigerian psychiatric profession is still strongly wedded to the biomedical model of the mind and primarily reliant on medication. There are also many stories of patients being psychologically abused, beaten and chained up. Mental-health patients are often forced to eat, sleep and shit in the same overcrowded spaces. Others have been left attached to beds all night whilst all the staff, due to underfunding, have gone home. Some patients as young as thirteen have been restrained overnight in buildings with no electricity, left with only a torch.[17] There are only eight federal psychiatric hospitals in the country.

Most patients end up being dealt with by religious authorities, mainly Christian or Muslim. This frequently leads to exorcisms. Otherwise, patients are left with their families. In 2020 a series of stories came to light about people with mental-health issues being imprisoned by those closest to them, usually due to the stigma attached to madness. Some were held in chicken coops or locked in box-rooms. Others were immobilised in gardens or garages, attached with rudimentary chains to large logs or machinery, sometimes for as long as thirty years.[18]

Many indigenous Nigerian cultures were rather open to treating and living with mental-health issues. Several Nigerian languages have long had explicit terms for different conditions and some areas, particularly in the southern part of the country, have a well-worn history of traditional remedies.[19] In contrast, many Nigerians now view mental illness as possession by evil spirits or the divine wrath of God. This, according to theologians and anthropologists, could be a hangover from when British colonisers translated indigenous religions from complex cosmological belief systems into simple demon worship. Those who prayed to false gods were not just heretical, they were mad. This depiction of 'insane natives' fused with descriptions of Africans as 'godless savages'. The spiritual forces of indigenous religions, percolated through tradition and culture, appeared anew following the introduction of monotheistic religions. This time, however, they were demons in a now Christian or Islamic world, demons in need of exorcising. This was initially a convenient way for the British to strip local populations of their humanity and undermine the very foundations of their existence. The fact that it threatened the stability of people's minds was not even an afterthought.

Resistance from Fela Kuti to Ken Saro-Wiwa

There is a long tradition of resistance in Nigeria. There is also a long tradition of violent government responses to resistance. The rebel musician Fela Kuti was famously on the run from the Nigerian authorities for much of his life. Whilst searching for Fela in 1978, the police threw his mother – Funmilayo Ransome-Kuti, a feminist activist and educator – out of a second-storey window, an event Fela immortalised in his song 'Coffin for Head of State'.

Ken Saro-Wiwa was another famous Nigerian dissident persecuted by his government. He was an Ogoni leader, writer and ecological activist who fought for the rights of the Niger Delta and its people. Saro-Wiwa was President of the Movement for the Survival of the Ogoni People (MOSOP), a nonviolent campaign to rid Ogoniland of the devastation wrought by oil companies, particularly Royal Dutch Shell (later Shell plc). Together, the movement wrote a bill of rights for the Ogoni people that was presented to the Nigerian government and the UN's Working Group on Indigenous Peoples, demanding political autonomy, economic justice in the form of restitution, a moratorium on oil flaring and dumping, as well as urgent medical aid.[20]

In 1993 MOSOP organised marches that drew together more than half of the Ogoni population, some 300,000 people. It made the news worldwide. The Ogoni, who had survived on the land for more than 500 years, found that after Shell began operations in the 1950s and legally started dumping its drilling waste straight into the Niger River, groundwater quickly became too polluted to drink, rainwater fell as acid rain, and oil slicks killed so many fish that fishing was no longer a realistic means of subsistence.[21] Teach a man to drill for oil and you fuck up an entire landbase. The Nigerian government decided to occupy the territory, militarily. Saro-Wiwa was devastated by the subsequent murder of 1,800 Ogoni people, 'in cold blood'. Many were raped, beaten and thrown in jail. Around 100,000 more were driven 'into the bushes and forests'.[22] Conflicts like this massively increase the likelihood and severity of depression, anxiety, PTSD, bipolar disorder, psychosis and schizophrenia. They also make it impossible for people to access care of any kind, which can easily turn a short-term acute situation into a long-term disability.[23]

In 1994 the military arrested Saro-Wiwa on completely spurious charges. He was held for over a year, during which time he was awarded the Right Livelihood Award, also known as the 'Alternative Nobel Prize'. In his acceptance speech, he declared: 'The inconveniences which I and the Ogoni suffer, the harassment, arrests, detention, even death itself are a proper price to pay for ending the nightmare of millions of people engulfed by the wasting storms of denigrating poverty on the sea of dehumanization.' He was tortured in prison whilst his region was ransacked. General Paul Okuntimo's troops had a mandate to 'sanitize Ogoni'.[24] Shell is on record as having paid Okuntimo's men, genocidal corruption that they tried to dress up as a 'show of gratitude'.[25] Eventually, after a show trial, Ken Saro-Wiwa was hanged from the neck until dead. Eight of his allies died with him on the gallows. Saro-Wiwa's last words were: 'I never destroyed my people, but my people are killed. If it be that I can die to free my people, please my God, and the Gods and soil of Ogoniland, allow me to die.'[26]

Shell no longer operates in Ogoniland but the area is still victim to oil slicks and run-off from unsealed boreholes and abandoned infrastructure. Shell is still very active in the Delta at large, spilling their rainbow sheen across water and land, and clogging the air with toxins and particulates. Six decades of oil extraction have left the Delta one of the most polluted places on earth.[27] As Niger Delta activist Saatah Nubari describes the situation: 'We have groundwater polluted with benzene 900 times above WHO level, we have farmlands with poor yields, rivers that are barely fishable, neonatal deaths numbering thousands yearly as a result of spills. We have reduced neuroplasticity of the brain as a result of oil pollution.'[28] In the first months of 2023, 13,000 people collectively sued Shell for the health impacts of existing oil spills, an important, partial win after years of work. Shell basically say that it wasn't them. The trial is set to be heard in the High Court in London in 2024. As well as the people, the area is also home to the continent's most extensive mangroves, which ecologists say are at risk of choking to death. According to the United Nations, clearing up existing spills and flare pollution would constitute the largest fossil-fuel clean-up in history, a gargantuan project set to take decades. Since the UN's announcement, more than ten years ago, work has only begun on around one tenth of sites.[29] That same decade has seen more than 8,500 additional spills.[30]

The 2022 floods hit the Delta hard. Given that this is a region knitted through by watercourses, thousands were forced to flee. The region's largest teaching hospital had to evacuate all of its patients, except those on the brink of death.[31] Shell offered a million dollars to flood victims in the Delta after heavy rains put 19,000 square kilometres under water. One million dollars equates to 25 cents per Delta resident. There is also an obvious and nasty irony to having a global fossil-fuel corporation offer a handout after extracting from and polluting your land, then selling what was beneath your feet on the international market so it can be burned, release carbon into the atmosphere and destabilise the climate to the extent that your home is flooded. When it was extracting and selling Nigerian oil, Shell tended to make an estimated 4 billion dollars per year from its activities in Nigeria; that is about 7 per cent of their total annual revenue. The Ogoni Bill of Rights demanded that local populations receive a fair share of the proceeds if these companies were going to continue operating. One million dollars is 0.025 per cent of what Shell made in a year from the Delta alone. At the same time, it is estimated that the copious amounts of fossil fuels that Shell extract worldwide are directly responsible for more than 1.5 per cent of total annual global emissions.[32]

Ken Henshaw founded We The People, a grassroots movement in the Niger Delta focusing on fossil-fuel extraction, climate justice and health. Ken has been working with Delta communities for eighteen years.

The reality is that climate change can be blamed for an increase in mental-health conditions in the Niger Delta region. It creates poverty and it creates destitution. When you have flood disasters and people have to move out of their homes for three months every year that's certainly going to lead to mental-health problems. People get depressed, they're frustrated. People become hopeless, so to speak. Largely because of the pollution and resulting deprivation, life expectancy in the Delta is forty-one, compared to fifty-six in the rest of Nigeria. That is bound to have serious psychological effects.

We also see an escalation of violence. That link clearly exists. The oil companies, the pollution, the poverty, the army – it all makes the Niger Delta a highly conflict-ridden place. That in itself is a cause of increased mental stress. If you're travelling on the river or the sea in an open boat with an outboard motor, at four or five different stops at least you'll be asked to raise your hands in surrender, in your own community. There's a huge impact of experiencing that powerlessness, that abuse. For the last two decades, the Niger Delta has been an area of constant conflict and fossil-fuel extraction has been the cause. When you live in a place where people are permanently fighting, when you're not sure that stepping out of your house you can return without being met by militants or the military, it definitely exerts mental pressure.

The militants are affected too. To be a militant you need to be seen as tough, to deaden your emotions. Huge numbers of young people are using drugs now, and that's

exclusively derived from the unique situation of the Delta.
The Delta is a warzone.

Since Saro-Wiwa's execution, armed operatives have taken centre
stage, many young men and women have been pushed to fight
after, at best, decades of neglect from the government and, at
worst, systematic oppression. Since 2006, eco-militants such as the
Movement for the Emancipation of the Niger Delta People
(MEND), the Niger Delta Avengers (NDAs) and the Niger Delta
Green Justice Mandate have all demanded ecological restoration,
economic justice for the region and deeper democracy. Their
demands have been backed consistently by threats: real threats.
Between 2006 and 2009, MEND, a 100,000-strong, fluid insurgency made up of loosely linked cells, nearly brought the Nigerian
economy to its knees. Equipped with camouflaged body armour,
speedboats and Kalashnikovs, they strategically sabotaged key parts
of the oil infrastructure to cripple operations.[33] National oil production dropped by 50 per cent.[34] The attacks largely ended when
then-president Yar'Adua offered generous cash payments, full
amnesty and jobs to 30,000 militants in return for a ceasefire.[35]

The ceasefire did not last, however. Many militants kept their
weapons, sceptical about the government's ultimate intentions.[36]
In 2016 there was another flare-up and two dozen major incidents drove oil production to a thirty-year low.[37] In May alone,
the Niger Delta Avengers blew up an underwater export pipeline
owned by Shell – a complicated mission involving deep-sea
divers – then destroyed a number of pipelines owned by
Chevron, followed by more owned by the Nigerian National
Petroleum Corporation.[38] The attacks continued throughout the
year. Nigeria lost its position as Africa's top oil producer. The
government ramped up its response, this time leaning heavily
on stick rather than carrot. This drove local communities further
towards the Avengers.[39] President Buhari deployed thousands
of troops, resulting in government-perpetrated bombings and a
shoot-on-sight order. The Avengers killed a number of soldiers
and took scores of oil workers hostage but, according to *Forbes*

magazine, the international press likely took little notice of the groups for the exact reason that they have 'shown little or no interest in maiming or murdering the innocent'.[40] The militants have a significant amount of support locally, often acting with the co-operation of Delta communities. In the eyes of sympathetic residents, according to the Council on Foreign Relations: '[the groups] agitate against the government for a better future. [These] agitations of the armed groups [are] an expression of their internal frustrations and yearnings.'[41] They have given out scholarships from funds earned from requisitioned oil, as well as paying for health centres and educational infrastructure the government has failed to provide.

Where mental health has been studied in the Delta, the prognosis is dire. Accessing healthcare in the region is difficult. The government has claimed that it is compensating communities for the impacts of oil companies through investments in 'social wellbeing'. Oil companies have also jumped in and thrown limited amounts of cash from their corporate social responsibility budgets in attempts to save face (and business). According to independent analysis, neither has made much of a dent. There have been no recognisable health improvements.[42] Meanwhile, the pollution continues, and so do the floods and the heatwaves. One small study looking at a women's mental-health clinic in the region found that more than half of those experiencing depression were classified as severe. The authors suggested that social stigma, discrimination and lack of education around depression were partly responsible for things becoming so bad before help was sought. But so was a manifest lack of access to support, which even when it existed, was shoddy and sometimes even persecutory.[43]

> The first thing [says Ken Henshaw] is for communities to recognise that these mental issues exist. You can gauge the humanity of a society by the way it treats people who have mental problems. In Nigeria, people see mental-health issues as people walking naked in the streets.[44] The government has to change this view, and admit that our system is

unequipped. They have to recognise that there are social drivers to mental health, real social drivers: how people generate income, the living situation, security, climate change, heatwaves, floods. These things all create conditions of mental instability.

Fear and lack of trust also play an important role, given the state's violent activity in the region. As Ken says: 'In the most affected regions in the Delta, there are definitely more people acting mad. The government responds by sending in the military. The government labels people rebelling against fossil fuels as "restive youth". They make it a cultural battle, instead of recognising that these people have been driven to mental breakdown by social, economic and environmental factors.'

The mental-health support that does exist is usually far away and people often lack the cash or the time to travel to access it. Those of us with mental-health issues, especially at the outset, are rarely the most enthusiastic when it comes to getting help. Every barrier put in our way can be just another reason to give up. The population of the area, including but not limited to the Ogoni, is 4 million people. They are served by just one functioning psychiatric hospital, whilst mental-health coverage is not provided in primary care clinics or local hospitals at all.

Inventing new worlds
These struggles are complex, with old, deep origins. So are the intricacies of people's mental states. Colonialism and capitalism play a part. So do indigeneity and democracy, culture and nature. Climate change is the new rough beast come to invite chaos and herald drastic change. The responses are not homogenous. Nor should they be. Resistance is never uniform or unanimously orchestrated. It is part of the process of inventing new worlds, a process that takes a plethora of movements. In Nigeria, at one end of the spectrum are the Delta Avengers who, in a way, constitute a pick-and-mix fusion of the philosophies of violent direct action advocated by Frantz Fanon and the popular mass movement for

Ogoni rights led by Ken Saro-Wiwa. Despite their relative popularity in the region and their efforts to rebuild their communities with reappropriated oil wealth, the actions of the militants are controversial. They sling rocks at oil Goliaths. They steal from the rich and give to the poor. But the use of violence is always controversial. I would argue that describing the Avengers' destruction of inanimate objects as violence is a sleight of hand long used by the oppressor to invalidate entirely justifiable methods of resistance. At the other more accepted and mutually supportive end of the spectrum is the work of people like Jennifer Uchendu and Ken Henshaw, organisations such as SustyVibes, TEAP, Mani and We The People. Some choose to fight underground (and underwater). Others choose to gather on the surface.

Even when they are focusing on tree planting or a litter clear-up, Jennifer's work opens up a space for people to link their chaotic internal worlds with the chaos outside. This can build on a sense of belonging to channel righteous anger into engaged action.

> There's something happening to our world that we don't have all the answers for. No one person can have all the answers. We have to make it as easy as possible to have a healthy balance of emotions, where we can also use the ugly and not-so-comfortable feelings, states of mind, as ways of building resilience. We should ultimately aspire for joy, hope, courage and care. But that can only happen in spaces where young people feel validated to speak about what they're experiencing. It needs to be okay to say: 'I feel so deeply about where my planet is going, what my future looks like, and I'm so depressed, angry and scared about my inability to make it change.' So, spaces that explore all of these emotions, they *allow* us to think through what moving forward could look like.

With the Nigerian mental-health infrastructure as poor as it is, Mani is the biggest provider of direct mental-health support for young people in the country. It is worth remembering that this is a non-profit, run by and for young people. They directly support around 1,000 people per month through their conversation cafés, run by some of their 1,500 part-time volunteers. In the past five years they have managed to train 35,000 individuals across Nigeria. They have an even wider reach across the continent. Their mental-health literacy programme is rolled out across virtually every social-media platform and in the last six months they had a reach of more than 350 million. They also provide free online therapy services. So does She Writes Woman, a charity run by and for Nigerian women with mental-health issues, set up by Hauwa Ojeifo following her survival of a suicide attempt. Nigeria has the third highest age-adjusted suicide rate for women in the world. She Writes Woman has a 24/7 toll-free helpline, teletherapy and hosts a digital community of like-minded 'mental-health enthusiasts'.[45]

Peer-support models like this are spreading across the globe, especially since the pandemic with its explosion of online meetings and online communities catalysed by isolation. In some areas, the number of people accessing online peer-to-peer support doubled after March 2020.[46] Radical mental-health activists have long been calling for patient-driven care, recognising the fact that those who live with psychological difficulties are *experts* in their conditions. Established researchers are starting to catch up, recognising that peer-to-peer support could be 'the future of mental-health care'.[47] As steeped as it is in misconceptions and assumptions, nearly a century of evidence from Alcoholics Anonymous and other 12-step fellowships proves that mutual aid amongst communities of sufferers can be transformative. The most successful movements of the 20th century, from anti-colonial struggles and the Civil Rights movement to votes for women and gay rights, all had sociality and communal care at their core. The Black Panthers set up health clinics and reading groups. The Suffragettes had tea rooms, protective spaces in which to confer, bond and feel safe. The Nigerian independence movement had highlife, a music and culture that carved out room for collective joy and 'set the mood of the Independence era by electrifying social space'.[48] Movements need culture, as well as spaces to support each other and build community.

What is needed, then, is not simply an expansion of mental-health provision in its more common forms but rather an open dialogue about what it means to suffer from psychological difficulties, ways to address this collectively, and the opportunity to explore the causes of and insights granted by these different experiences of consciousness. We need to celebrate diversity in all its forms, and that includes an array of approaches to defining mental health in and of itself. Incendiary appeals have been made directly to the psychiatric community for decades for the inclusion of social, cultural, economic, historical and ecological factors as causes and remedies of mental-health issues. Systemic problems require systemic solutions. 'I was always very curious,' Jennifer tells me, 'to understand the role of power, colonisation, oppression and extractivism, in what we now call everyday climate change and climate crisis. It became overwhelming. I crashed. But I'm grateful for that. It's the anxiety and rage that led me to speaking to you about this now.'

We need something closer to liberation psychology, a thread in critical psychology now growing in prominence in climate circles. Liberation psychology is a way of interpreting and responding to psychological challenges, inspired by communal practices and rituals that have been present in countless forms for millennia, but pulled together and distilled in modernity by the Salvadoran rebel psychologist Ignacio Martín-Baró, also a scholar and a priest. He wrote his masterpieces in the heat of the Salvadoran civil war, before being murdered by the army in a university building, along with seven others, in 1989.

> Psychology must switch focus from itself, [Martín-Baró wrote] stop being preoccupied with its scientific and social status and self-define as an effective service for the needs of the numerous majority . . . The primary object of its work [is that] of serving the need for liberation.

The truth of the ... people is not to be found in its oppressed present, but in its tomorrow of freedom; *the truth of the numerous majority is not to be found but to be made.* So to acquire new psychological knowledge it is not enough that we base ourselves in the perspective of the people; it is necessary to involve ourselves in a new praxis, an activity that transforms reality, allowing us to know it not just in what it is but in what it is not, so thereby we can try to shift it towards what it *should be* [emphases added].[49]

Liberation psychology looks at the systemic, social and environmental causes of mental distress, rather than piling blame onto the individual. By encouraging people to see their mental health as intimately connected to the state of the world, Jennifer and her team actively collectivise people's experiences. At this point, 'we're all in this together' becomes more than just a slogan. To return to Fanon – whom some see as the first liberation psychologist in Africa, although the term was not coined until more than thirty years after his death – he believed in action, not just words and medication, as a means of transforming the mind and transforming society.

'We revolt,' he wrote, 'simply because, for a variety of reasons, we can no longer breathe.' This quote was plastered all over the streets at the height of the Black Lives Matter movement, connected to George Floyd's brutal suffocation and famous last words. Fanon campaigned for the reform of asylums and the revolutionising of psychological treatments. One treatment method Fanon championed was socio-therapy, which integrated aspects of the patient's cultural and historical context with the society in which they lived, whilst bringing in elements of how people define the world from their position. For him, acting in and on the world is a vital part of maintaining freedom and sanity.

A friend of mine who works with families as a social worker tells me how this space, too, is crippled by the belief that just talking and thinking things through is enough.

'It's frustrating to work with the perception that what's needed is assessing, identifying and then "talking through the problem".

That's an important step, but I feel like my job has always been to say "then what?" Because what we're trying to do is create change. Change doesn't just happen by thinking of how you got there or what's wrong (like, "the past sucked and it was really hard") but concrete practices, concrete ideas the families usually already have. That can include talking about the past in a different way, but every conversation needs to start with what change do you actually want to happen? That kind of work becomes about collaborating and making a strategy. It's about doing something.' Whether or not they reach this voluntary but break-through step can be the difference between a parent keeping their children or losing them. There is a similar dynamic in most therapy.

Personally, recovery coaching and practical changes to the structure of my life have had a far more significant and lasting impact than conventional talking therapies and medication alone. Activism has done the same. In the case of climate-related mental-health difficulties, consciously directing some of these actions towards the systemic causes of suffering is fundamental. This is true if it is in the context of activities that are as extreme and underground as the Niger Delta Avengers, or as open, communal and supportive as the work of Jennifer Uchendu, Ken Henshaw and Mani. Fighting for a better future can be a form of recovery.

'Talking therapy is important,' Jennifer tells me, 'but it needs to be converted. There needs to be some kind of next step after that conversation.' Virtually every successful trauma therapy involves revisiting and establishing a different relationship with the root trauma. Climate action can do that for climate-induced mental-health issues.

'That's exactly what we've been doing,' Jennifer tells me. 'To give one example, we talk a lot about the flooding in Lagos. People don't really think they can do anything about it, but that's not true. We are by the sea so sea-level rise impacts Lagos drastically. There's a massive drainage network issue, too, and plastic pollution and litter culture that clog up drains and other routes that could let water escape.'

In 2022, Lagos saw flooding that affected nearly 25,000 homes. The papers were filled with pictures of submerged cars and stories of families sleeping on floating 'slabs' they had made in flooded rooms.[50]

> We take people to the streets [says Jennifer]. Young people have 'street conferences,' conversations with the public about the impact of that one bottle you threw out, they point out the mess of plastic in the gutters. Psychologically, it's taking your life in your own hands: creating dialogues, education, building relationships. This reconnects people to nature and to each other. That sense of empowerment and agency can be helpful, you know, really, really helpful. This year we had a massive, massive, massive turnout of young people from different parts of the country, people within our community saying 'I want to help'. I mean, it doesn't particularly mean that you won't wake up with intense climate emotions, but it changes the game.

As well as combating climate change and supporting people's mental health, these projects are also experiments in radical democracy. Each and every new person is a repository of information and ideas, a unique mind to add to the overall potential of the movement. We don't know where it is going but, as Jennifer reminded me, we don't have all the answers. Nor should anyone at this point.

Wellbeing is part of the fight
We need a proliferation of mutually reinforcing groups, pushed forwards in part by a promise of wellbeing in the face of many dangers. This wellbeing needs to be part of the fight. We need to work, too, on being kind to ourselves. For Jennifer, a crucial turning point came when she chose to let go of whether she was 'doing enough', and instead started asking herself whether she was 'doing her best'. 'That helps me to shift perspectives,' she tells me. 'There's so much to be done, and there's only so much

I can do as a person. When I move to 'doing my best', I'll still really put in grit, energy and passion, spending nights trying to organise with people so that we can take action at a community level. Absolutely. That helps me sleep better at night and that helps me, you know, regulate emotions as much as I can. It helps me to understand that life is to be enjoyed just as much as it's really difficult, so I can find joy in being alive.'

One would hope that the recently signed Mental Health Act would do away with the need for volunteers and charities to save people's lives by granting access to basic mental-health support. The Act's authors promise that it will give human rights to mental-health patients, legally defending them against discrimination for the first time, whilst also giving them the right to participate in their own health plans. The seclusion and restraint used in hospitals will also, we hope, be outlawed.[51] That is obviously a step in the right direction and the extraordinary efforts of those who got the Act this far should be applauded. It should also help to destigmatise mental health in Nigeria and hopefully improve people's understanding of and compassion towards sufferers which, as we have seen, is key. But without a far-reaching education plan and drastic changes to the whole philosophy of providing mental-health care, let alone the underlying systemic drivers, it cannot possibly solve the present mental-health crisis – let alone the ones on the horizon. To save ourselves, we have to go deeper. We have to rebel.

5

Pakistan: planting resilience

'That day, I sensed Pakistan's tree to be ... greener than before and fresh new leaves coming up from thin branches. I kept reciting *La hawla wala quwwata* [there is no power but Allah]. I tried to control myself but all in vain. There was a mysterious feeling in this magical world. A strange intoxicance. My senses tried to control my mind but I was lost in the magic.'

Mumtaz Mufti, *Piyaz ke Chilke*, 1968

WATER DEMAND IS ROCKETING in India and Pakistan, just as reliable supplies evaporate. Each nation controls watercourses that pass through the other's territory, agreements made on then-valid assumptions that average flows were a reliable constant. Pakistan has absolute rights to the Indus, Chenab and Jhelum, whilst India is entitled to all the water in the Sutlej, Ravi and Beas. These agreements have been broken, repeatedly. The nations now face a toxic brew of climate-driven water scarcity and state machismo, underwritten by both countries being nuclear powers.[1]

Pakistan is home to more than 7,000 glaciers. Collectively, they constitute the largest bodies of glacial ice in the world outside the Arctic and Antarctic. Almost three-quarters of the runoff from the upper Indus, Pakistan's principal water source, comes from glacier melt, but the IPCC predicts that the Himalayan glaciers feeding the subcontinent's six main tributaries could shrink by 50 per cent by 2050.[2] Ice-melt will feed into rivers and floodplains with unpredictable intensity and frequency, resulting in oscillations between drought and flooding.[3]

Pakistan is home to almost 3 per cent of the total global population.[4] It is the fifth most vulnerable country to climate impacts, despite being responsible for just 0.3 per cent of total global cumulative emissions.[5] In 2022, Pakistan saw a heatwave of unprecedented proportions. In April, the city of Jacobabad hit 49 degrees Celsius, one of the highest temperatures ever recorded on earth for that time of year. Sherry Rehman, Pakistan's minister for climate, described the situation as an 'existential crisis'.[6] Some 200,000 people were put at severe risk by the drought, the worst on record in some regions as livestock died, water sources evaporated and thousands were forced to migrate.[7] Scientists say droughts like this are already thirty times more likely because of climate change.

What came next was far worse. The climate minister's prognosis evolved from the 'existential crisis' of the droughts, to brute 'dystopia'.[8] As the monsoon season hit, torrential flooding inundated the scorched ground. Some described it as the end of days. For many, that would be a statement of fact. A third of the country was submerged. In some areas the depth was measured in fathoms. The rivers and streams, already bulging from climate-induced glacial melt, were further filled by unusually torrential monsoon rains. Rainwater flashed off drought-hardened soil and the Indus burst its banks. Over a million homes, 130 bridges and 13,000 kilometres of roads were swept away by shelves of water carving across the landscape.[9] Burst sewage mixed with the water, and disease spread. Those who could, got into boats. Videos recorded on smartphones by those lucky enough to escape the torrents show entire buildings yawning apart, tarmacked highways bucking into twisted knots, and pylons, fences, cars and people careening down thick coils of water, muddy-brown and writhing like huge, demented pythons.

Almost 2,000 people were killed. Over 12,000 were injured. An astonishing 6.5 million were left in dire need of help.[10] In total, 33 million people were directly affected, around 15 per cent of Pakistan's entire population. These kinds of numbers are difficult, if not impossible, to credit. Looking past the incomprehensible numbers, into the actual lived experience of those who have had their lives eviscerated, is even harder.

Dr Asma Humayun is a Pakistani psychiatrist who has been working to develop mental health and psychosocial support services in the country for more than two decades. She shared the story of a man who had to make an impossible choice, without a moment's hesitation. He saw one of his children being ripped away by floodwater. His cow, in the same instant, was also at risk of drowning. He had to decide, in a split second, whether to save his child, or save the cow. That might sound like an easy decision but the father knew that the cow would be able to feed his six remaining children in the aftermath of the disaster. These kinds of choices, regardless of the decision made, leave deep scars. It was a deep, definitive fork in the road that will stay embedded in this man's psyche for life.

The floods have overwhelmingly impacted the poorest. The poor are already the least likely to receive mental-health support of any kind. There is a high prevalence of mental-health disorders in rural communities in Pakistan with 44 per cent of men and 72 per cent of women in the countryside reportedly suffering from some kind of psychological distress, a far higher proportion of the population than in other countries of similar socio-economic standing.[11] The mean overall prevalence of anxiety and depressive disorders alone is 34 per cent.[12] Poverty and lack of education are part of the reason. For women, the heavy weight of a rigid patriarchy is also a massive factor. It was only in 2019 that the Senate made it illegal to marry women off before the age of eighteen. Still, in 2021, nearly one in five women aged twenty to twenty-four had been made to marry as children.[13] Domestic abuse is scarily high. A quarter have experienced intimate partner violence and sexual abuse, nearly 15 per cent in just the last year.[14]

'If you did a survey of all the people most affected by the floods, I'd say one in a million would blame climate change,' says Fahad Rizwan, a climate activist and reformer in Pakistan. 'The rest,' he tells me, 'will put it down to the wrath of God.'

This does not mean that people have failed to notice a change in the climate. Indeed, those working the land are often far more highly attuned to such changes than almost everyone else, but causal attribution is another question entirely. In a society as diverse in sect and yet almost uniformly Muslim as Pakistan (94.5 per cent of the population practises Islam in some form), similar spiritual causes are given to the understanding of mental-health issues. Rural Pakistanis are more likely to attribute conditions like schizophrenia to 'God's will' or what academic scholars call 'superstitious ideas', than to see them as medical disorders.[15] Even in urban centres like Karachi, more than half of people see depression as a natural state of sadness, rather than anything explicit. Similar numbers believe that praying to God is amongst the best treatments for mental-health issues. Before those in the Global North raise their eyebrows, it is worth noting that around half of evangelical Christians in the US think that depression can be prayed away. Western scientific studies also seem to suggest that spiritual practices, broadly defined, can significantly benefit people with mental-health issues – even those who strongly identify as atheists.[16]

Psychiatric hospitals existed in the Islamic world long before the British came to what is now Pakistan.[17] In fact, the first psychiatric hospitals in the world were built by Muslim societies, as early as the eighth century in places like Baghdad, Fes and Cairo.[18] But the mental-health support that exists in Pakistan today – academically, physically and culturally – is primarily a result of British colonialism. By some accounts, there was a brief period when British institutions were relatively communal. Understaffed and under-resourced 'asylums', as they were known then, even utilised local community knowledge to provide relative refuges for the 'mentally ill', but from the mid-19th century onwards Western medics were sent in and the hospitals were predominantly used by the British as one of their many means of forcing subservience onto domestic populations.[19]

Since the end of the 18th century, the British had used 'native-only lunatic asylums' in British India to thwart cultural and political opposition. These had begun as crude, profit-driven spaces used to force social outcasts into free labour and morphed into tools of social and cultural control that promoted Victorian ideals of individual betterment and civility.[20] By the era of Partition, these units were not only used for cultural realignment, but weaponised as effective ways to pathologise and dehumanise those resisting British rule.[21]

An absence of mental-health care

'Mental health has never been a priority on the country's health agenda,' says Dr Humayun. Even today, just 0.4 per cent of the national government budget is dedicated to mental health. 'At Partition,' she tells me, 'we inherited three long-term psychiatric hospitals, because of where the borders were drawn. It wasn't until the 1970s that enough Pakistanis trained as psychiatrists to set up tertiary care hospitals. Psychiatric services have grown around psychiatrists, and the available care is concentrated in urban centres, with a very biomedical approach.'

It is estimated that 80–90 per cent of those suffering from mental-health issues in Pakistan have no access to treatment.[22] This number is probably far higher due to a lack of consistent reporting, just as the number of dead from the floods is likely to be much higher than we know. Around 225 million people live in Pakistan. That is equivalent to around 70 per cent of the population of the United States. There are around 500 psychologists and 400 psychiatrists in Pakistan;[23] the US, by comparison, has around 45,000 psychiatrists. If we take the widely used estimate that one in four Pakistanis are struggling with their mental health, from depression and anxiety to psychosis and bipolar disorder, that means each individual psychiatrist is responsible for the care of around 140,000 patients. To give each of these patients a single one-hour session would take forty-eight years of working eight-hour days, seven days a week.

Of the eighty districts most affected by recent flooding, fifty-five do not have a single psychiatrist. Dr Humayun has long been working to implement alternative, more humane and sustainable models of mental-health care in her country. When the flooding hit, her work became all the more urgent.

> To understand how the floods affect mental health in Pakistan, [says Dr Humayun] you need to take a step back and understand people's experiences of security and insecurity before the floods. In London, your security might revolve around buildings and systems. But in rural Pakistan, the social security is the land, the mountains and the environment people live in, relate to and feel secure in. All of a sudden there has been total devastation. It includes their houses, their animals, their land. They are left with nothing. For months people are sleeping out under the sky. Vulnerable people are likely to resort to drugs. Women and children are likely to be subjected to violence and abuse. Children who were already malnourished are now dying from starvation. When homes were washed away, many people piled their belongings in the corners of demolished houses. Others would come in the night to steal utensils, clothes, anything. I think the biggest impact on mental health, is the destruction of hope. What are people supposed to hang on to?

As an adviser to the Ministry of Planning, Development and Reforms, Dr Humayun developed an innovative, evidence-driven and scalable model which can help address huge treatment gaps. The Mental Health and Psychosocial Support (MHPSS) model goes beyond the standard biomedical model, incorporating different theories of mind, looking at various possible causes for mental ill-health – including stress, life events, physical ailments, views about the world, and more – as well as putting human interaction at the heart of treatment models. MHPSS, as Dr Humayun is pioneering it, draws on the International Classification of Diseases (ICD), a WHO-certified global alternative to the psychiatric reference document most widely used in the Global North, the infamous *Diagnostic and Statistical Manual of Mental Disorders* (DSM). Dr Humayun was part of drafting the ICD section on stress-related disorders. The ICD is adapted for various cultures in the wider world and focuses on distress and trauma related to adversity in life, relegating the narrow Western reliance on brain-glitch theory and pharmacology. Providing psychological support in humanitarian zones and elsewhere, Dr Humayun says the most important thing is to identify the normal responses to distress; building on people's inherent ability to cope; and intervening psychotherapeutically to prevent complications that hamper recovery. This is about building resilience for recovery, not just putting gaffer tape on a fractured dam.

Over recent years several hundred Pakistani practitioners have been trained in identifying trauma, particularly focusing on eye movement desensitization and reprocessing (EMDR): a PTSD treatment using eye movement and memory reprocessing. The risk is that now PTSD cases might start appearing across the country, especially in hotspots where those practitioners are active. But, as the old saying goes, if all you have is a hammer, everything looks like a nail. In the same way, just flying psychiatric medications into disaster zones is never going to plug real gaps in human need.

Dr Humayun tells me that, according to the WHO, at least one in five people is likely to suffer from a mental-health condition in a humanitarian crisis. But, she cautioned, just assuming that the WHO's estimate is correct and simply taking the corresponding number of antidepressants into an area is a mistake. Medication alone cannot solve what is at the root. It's important not to medicalise distress, she tells me, and there is a really urgent need for psychosocial support. 'I'm going in there to talk to people and see how they're coping, how they're recovering with time, where they are on the distress spectrum.' It is only through human contact that it is possible to provide meaningful support.

This bridges an important gap between seemingly sacrosanct Western definitions of mental disorders and the internal perceptions of those experiencing them. Western medicine deifies objectivity – or at least the appearance of objectivity. That might sound like a ridiculous statement. The scientific method has indeed furthered much understanding of the human body and mind in recent centuries. But the lived experience of ill health and divergence, whether physical or mental, is first and foremost phenomenological. Experience is embedded in cultural realities, as well as material ones, including but not limited to the climate. Psychologists, therapists and psychiatrists all too often tend to try shepherding largely subjective, qualitative experiences into neat categorical boxes using quantitative scales that then allow for relatively uniform treatment.

In some of the towns in northern Pakistan where Fahad Rizwan works, 'very religious men,' he tells me, 'would see depression as a form of spiritual weakness'. Using a standard diagnostic questionnaire reliant on placing oneself on a subjective scale, like the Beck Depression Inventory, to determine whether or not someone should be categorised as unwell is almost laughable. If religion encourages people to be more stoic, results could be skewed to the point of irrelevance. One of the Beck questions, for instance, refers to 'feeling like a failure', whilst others ask about feeling sad, crying and sex. The internally interpreted experiences of these, as well as what it is possible to communicate, are deeply tied to an individual's cultural situation. To a large extent there is a circular logic defining psychiatric conditions within clinical boundaries, then using those clinical boundaries to define conditions, and so on. Messy as it may appear, what we need is a more discursive, relational and systemic approach to mental health. It would be more subjective, but also more useful and adaptable to an era of profound, divergent psychological upheaval.

The ICD now defines a condition called 'adjustment disorder'. Adjustment disorder often appears in the aftermath of catastrophes like Pakistan's floods. Sometimes people can psychologically recover from it on their own. Other times the distress persists, lingering like a spectre of discontent, tragedy or madness at an individual, family or community level. In Pakistan, as the flood waters recede, people's levels of distress should gradually drop accordingly, even if their former lives never truly return and their lands take decades to recover. But if people are still suffering mental ill health to the same degree after six months or so, then they will be primed to suffer other conditions such as depression or anxiety.

In some ways, these kinds of catastrophes can lead to a step change in health systems. Many people who were suffering from pre-existing conditions, but were not reported or treated, might get noticed once their condition is exacerbated by the floods. 'There is always opportunity in a crisis,' Dr Humayun tells me. 'People suffering with depression were previously just sitting there untreated. At least now they can begin to reach for help.'

The parents of one thirteen-year-old girl, for instance, reached out to Dr Humayun's MHPSS programme saying that their daughter's behaviour meant that they needed to chain her to her bed so she would not run away from home. Doctors from seven hours away were able to see from video footage that the girl likely had a learning disability and compounding mental-health issues. Before the floods she was untreated. The floods accentuated her behaviour and the family's capacity to help her diminished. 'Now that she's in the system, following the floods, there is an opportunity to help her and lots of other children with learning disabilities and mental-health issues. We can give guidance to the parents and help build systems of education. There is an opportunity here.'

One condition recently added to the ICD, prolonged grief, is a severe illness which is also recognised in the DSM. Prolonged grief is very likely to be a common result of the devastation in Pakistan. In Sindh, one of the worst affected provinces, one woman was stuck in hospital wailing and crying. People assumed she was mad and left her alone. Eventually someone came to ask her what was wrong. She needed 1,000 rupees (less than $4) to bury her child who had died in the floods. As soon as somebody gave her the money, she stopped screaming, got up, and left to organise the burial so she could then look after her remaining children.

A similar story involved a father who, lacking the money for a burial, kept the body of his dead daughter in the family tent.

This was the same tent where he slept with his wife and surviving children. The family sat in the enclosed space, with the dead body, waiting for the means to bury her. Anything that complicates the process of mourning and grieving massively increases the risk of prolonged grief, an abnormally extended and skewed state of despair, also characterised by lethargy, anhedonia (the reduced capacity to experience pleasure) and loss of meaning and purpose. Losing a child is always deeply traumatic. Many say there is no worse experience. But losing a child and then having to sit next to the corpse day after day, night after night, is going to leave a scar so deep it could tear you from reality entirely. These warped psychological states have the potential to cascade down through generations. Being raised by a parent so bereft, perhaps only intermittently able to provide care, can seriously damage a child and their relationship with the world.

Most of these stories are untold. The level of humanitarian support is horrifically low in comparison to the level of need. A huge proportion of the population most affected by the floods have not been contacted by the government or NGOs. We are often aware of this when it comes to physical needs like shelter and clean water, but rarely do we think of mental-health support as a condition for humans to flourish despite how fundamental it is. Dr Humayun, amongst others, describes this as dehumanising. 'If you are in a healthcare system that's mostly biomedical [as Pakistan's is], you ignore distress until somebody needs a pill. That is also dehumanising.' Purely from an efficiency standpoint this makes no sense, anyway. Prevention, as we know, is better than cure. This does not mean that everyone needs a mental-health medical team, but at any point on the distress spectrum people can at the very least be listened to.

A groundbreaking new model

Dr Humayun is now running a large-scale MHPSS pilot in Islamabad, with the hope of rolling it out across flood-affected areas as soon as it is ready. With the support of the Ministry of Planning, Development and Reforms, it took Dr Humayun and her team a month to train 1,000 people in psychological first aid and conduct sensitisation training to overcome discrimination. These 1,000 were largely picked based on their position in the community, prioritising those with higher social capital like teachers, rescue workers and police officers, who also interact with many different people every day. Newly trained to notice potential symptoms of distress, this group – known as the 'Hamdard Force' or 'Compassion Force' – communicate with the rest of the team using a mobile app. The pilot is tiered, so the Hamdard Force can also talk to clinical psychologists, primary-care physicians, mental-health specialists and consultant psychiatrists for supervision. The app, vitally, does not require fast internet.

The very simple design of this pilot allows 1,000 people in the community to work as the 'eyes and ears' for mental-health specialists and to connect individuals, with their consent, to practitioners who can monitor their health and may be able to help them. There is a button for volunteers to ask for support plus a helpline, built in a way that allows practitioners to call back rather than offer twenty-four-hour support, allowing for a small, adaptable and effective team. More than half of those referred during the testing phase ended up speaking to the team. The network is also a means of distributing educational material, enabling the team to offer further training and quickly generate widespread understanding of mental-health issues. The programme can easily cover a population of around one million, which is the average size of a district in Pakistan, with the support of just a few specialists. In the testing area there was previously only one psychiatrist.

The model could easily be scaled up – applied more widely throughout the country and the world – and the resources needed to maintain it are extremely low. An early iteration of this system for an internally displaced population in Bannu, Pakistan, found that more than a fifth of those referred to it came back voluntarily to follow-up, a surprisingly high figure. Many of the patients came from indigenous communities. Most spoke only Pashtun.

Bannu, a city in the Kurram district bordering Afghanistan, was the site of much conflict in recent years and had an extremely high level of psychological distress, even when economic factors are accounted for. In 2009, 3 million people were displaced by Taliban conflict. In Bannu, at the time of Dr Humayun's study, the population of 1 million was, again, covered by just one practising psychiatrist.[24]

Fahad Rizwan and his Green Squad team are also working in Kurram district. The population is beset by intermittent Sunni–Shia conflict, patriarchal exploitation and abuse, poverty, climate shocks and a severe lack of medical infrastructure. In a recent workshop Fahad organised in the community, a group of local men had a discussion facilitated by a psychiatrist. They found that climate change was leading to increased levels of distress, intimate partner violence and children running away from home. When they dug a little deeper, the team found a thick fabric of causality lying beneath the community's psychological struggles. The forty-year conflict raging in the region was central. The conflict was itself largely fuelled by deforestation and resource depletion. What was previously old collective land, managed by a tribal system, had been broken by colonialism then unravelled in a tragedy of privatisation. Levels of territorial aggression between Sunni and Shia communities are still high. Women are deeply subjugated in many of the villages. 'Mental-health issues,' Fahad tells me, 'from depression, anxiety and loneliness, to PTSD, mood swings, or psychosis ... it's all seen as a "lack of spirituality". People suffering in this way are just seen as being very far away from God.'

Fahad first came to this work through tree planting. Ten years ago, he started planting saplings under his own steam. 'The streets were so hot I was burning my feet on the way to school,' he says. 'I originally planted them for shade.' As he met other activists, Fahad quickly realised the web of connections between tree planting and climate change, but also the broken, corrupt political and economic system of Pakistan, as well as its militantly religious cultural norms. Planting trees was direct action. It was a rupture. Unilateral afforestation was a reversal of the expectation to stay in your lane, to respect hierarchical authority and refrain from subverting tradition.

Soon, the authorities began to see Fahad as suspicious. They called him in for questioning. First the government, then the military. Islamist militants threatened him. They labelled Fahad an agent of the West, preaching values like liberalism, women's rights, sensitivity to LGBTQ issues and the like. He pulls up his lip as we are speaking and he shows me his two front teeth, which the Islamists broke in a violent attack. 'They thought I was an American spy,' he tells me. Soon, Fahad figured out a workaround. 'If you bring capitalism, political economy and climate change into it,' he said, 'you get into huge trouble.' It was the wider analytic lens the authorities had a problem with. If, instead, he told them he was simply planting trees to increase biodiversity 'for the birds and the bees' then the authorities would leave him in relative peace.

A decade later, Fahad and his now 700-strong team are still planting. The work has been hard and it has necessitated a great deal of social and psychological support for those involved. They organise restorative nature walks, do group yoga and connect people with help when they need it. 'Everyone's scared,' he says. 'But when people start to engage, when they learn more about climate change and alternative ways of living, they become part of the collective mission to save the planet. They have hope. They have purpose. Religion is like this. It provides something people need: belonging. Human existence needs a mission larger than ourselves, and climate change is big enough to be that.'

Stuck within the guardrails of the country's dominant ideology, Fahad and his team soon hit a wall. Even when they had reforested whole areas, the saplings were often destroyed. One of Green Squad's members, Abdullah, passed by a grove of trees he had planted in Karachi seven years before. Every single tree had been cut down. He plunged into a trauma depression. Abdullah does not plant trees anymore. 'The planting drives really hurt – there's something that just breaks inside.' Abdullah didn't leave the movement, though. He is doing a lot better, writing and managing Green Squad's burgeoning social-media accounts. Similar blows have knocked the wind out of Fahad, too. 'Still,' he says, 'I always try to give people hope. I know there are people having far more difficult struggles, and deep down I am a man of nature. Doing this work brings me peace.'

When I asked Fahad whether he felt like they were sometimes fighting a losing battle, he gave me a flat 'yes'. Pakistan is rapidly urbanising and new infrastructure means the indiscriminate razing of habitats all over the country. 'It's a very colonial kind of capitalism,' he says. 'There's no planning. It's extractive and it's destructive.' Corruption is systemic. Much of the money given to the Pakistani government by the UN ends up in the pockets of the elite. A recent grant that Green Squad won to plant hundreds of thousands more trees got snarled up in government bureaucracy. They expected a 10-per-cent kickback for a fictitious 'planting certificate'. 'Even the reparations money or the loss-and-damage funds,' Fahad says, 'virtually none of that will reach the people that actually need it.' After a pause, he tells me he can say these things in English. Never in Urdu.

If you listen to the international press and the congratulatory, back-slapping rhetoric of the World Economic Forum, you would be forgiven for thinking that the Pakistani government has caught the reforestation bug, too. Pakistan's previous prime minister introduced the 'Ten Billion Tree Tsunami Program'. Its aim is to plant at least one tree for every person on earth. To date, according to the government, three billion trees have been planted and half a million jobs have been created.

Planting trees is an important step, but some say the popularity of the Pakistani government's programme is largely down to its simplicity. This is also its weak point. Not every sapling will become a mature tree, especially if they are unattended. Many carbon-offsetting companies take advantage of this problem, advertising and taking money for net carbon reductions equivalent to a mature tree's lifetime, even though this will take decades to achieve. The sapling might just wither and die. Experts argue that to be effective carbon sinks, a typical forest must mature to one hundred years. Even then the wood needs to be put to long-term use, rather than be burned or allowed to rot. Saplings planted to extend forest cover absorb around a fifth of the carbon that more mature trees do. There are a host of other potential pitfalls, too. The 2022 California wildfires, to give just one example, reduced whole swathes of carbon-offset reforestation to ashes and, importantly, to carbon dioxide.[25] Planting new trees to replace the old ones, like for like, is a fool's errand. New plantations are often mono-cultures arranged in rows, closer to supermarket aisles than the dense biodiverse thickets of mature ecosystems.

It is better than nothing, but when I asked Fahad about the Ten Billion Tree programme he said: 'This is the reality, Pakistan's deforestation rate is extremely high. Our forest cover has shrunk massively over the years and we are nowhere close to recovering.' Pakistan still has the second-highest deforestation rate in all of Asia, behind Afghanistan. Annually, around 27,000 hectares are destroyed. The country's recommended forest cover is 25 per cent; currently official reports claim that it stands at around 5 per cent.[26] This is a disputed figure. Fahad tells me the most recent reliable data puts it at closer to 2 per cent.[27]

Most people living in rural areas depend on the forests for firewood, which is one of the largest drivers of forest loss, driven as they are to overlap and share smaller spaces. Logging and urbanisation are also significant contributors. In recent years, wildfires in northern parts of Pakistan have emitted decades' worth of stored carbon in days. Historically, these trees absorbed floodwater, maintained soil integrity and acted as physical flood barriers. Now loggers, some of them operating illegally in indigenous territories, have been known to use rivers as a means of transporting felled trees. During the floods these became battering rams. Where once they would have calmed the flow, trees are now riding the waves and smashing their way through infrastructure, savaging landscapes and clogging up waterways.

It cannot just be about birds, bees and trees, vital as those are. In her short book *Walking with the Comrades,* Arundhati

Roy writes about a tribal community fighting corporate interests on the border between India and Pakistan. Government forces, companies and paramilitary units are trying to privatise the forests and evict the people living there. The real enemy is capital. The community's entire way of life is being threatened because dwelling in that landscape without explicit property rights is irreconcilable with the two nations' socio-economic systems. In 2020, *Walking with the Comrades* was banned by an Indian university for being too radical. Roy was, she said, 'not in the least bit shocked or surprised.'

Tree planting and mental-health co-ops

Fahad is still planting trees, but he is also planting the roots of alternative economic structures in some of Pakistan's peripheries. This involves ecology but is also about mental health and climate resilience – as well as peacebuilding, economic security, feminism, redistributing power and social equity. Green Squad is introducing tree planting and agro-ecology co-operatives to Kurram district. As well as being criss-crossed by conflict, the area is also a hotspot for suicides linked to crop failure, as well as climate-related depression, PTSD, domestic abuse and violence linked to drought, heatstroke and flooding. Some people have even been driven to radicalisation, Fahad tells me. There is a strong correlation between the Pashtun and Punjabi regions most affected by climate change, recent rises in hospitalisations for mental-health issues and increased rates of religious militancy.

First, to ease in the new ideas, only male villagers were involved in the co-operatives. Then they combined Sunni and Shia groups, followed by the introduction of women to the workforce. As co-owners of the enterprise, these people were all given a voice and the social infrastructure to communicate and operate on a far more equal footing, tentatively at first, then habitually. In some cases the women involved had never left the house before. Since the establishment of the co-ops, however, in Fahad's words, they have collectively 'broken the patriarchal structures, proven that they're obsolete.'

Many religious skirmishes have flared up in the region since the programme's inception. Not a single one has broken out in the areas where these co-operatives are running. 'People who previously couldn't shake hands,' Fahad says, 'are now going to each other's weddings, funerals, religious ceremonies.' Economic goals and a focus on wellbeing were key to getting the local mullahs and clergymen on board. If they had simply approached these authority figures with the idea of planting trees, few would have participated. Some still frown on the project, particularly some elders, but overall, the co-operatives are popular in the villages.

The co-operatives draw on both traditional and modern skills to create a different, dynamic relationship to each other and the land. People make decisions together. Grazing patterns have changed. Cattle and goats are no longer stunting trees, as they are now kept in pastures modelled on Israeli and Australian systems. Crop diversification is increasing resilience. More biodiversity means that people can start to use the forests as a commons, where once they were sites of conflict over scarce resources, and even people trafficking, abuse and abandoned children. These new social and economic institutions help people remain more rooted. Or, as Fahad puts it: 'People want to stay.'

Today, lots of governments see bioregional localism as out of date. Climate shocks are caused by global drivers. But projects like Green Squad's Kurram co-operatives are powerful examples of how local knowledge can result in an adapted way of living, one that prioritises ecology and wellbeing without any need for government intervention. It could be the first in a wave of decentralised alternative structures that address both resilience to climate change and elements of the mental-health epidemic. What is happening in Kurram district is a collective restructuring of social relations. Communities are healthier, happier, richer, and more prepared for climate shocks and their unwelcome bedfellows. It is worth remembering Dr Elaine Flores' work suggesting that one of the strongest predictors of how well a group of people are likely to cope with a natural disaster is neither GDP per capita nor the quality of emergency services, but how densely woven the community fabric is prior to the shock.

By that metric, Pakistan as a whole is currently doing badly. Local communities may be well knit, but national inequality is searing. Emblematically, Pakistani Prime Minister Shehbaz Sharif co-owns a multimillion-dollar steel conglomerate. He was imprisoned for seven months for money laundering just prior to his inauguration. Since the 1990s, free-market reforms have meant that income growth in Pakistan has mostly benefited the rich, whilst large amounts of wealth have been exported outside the country. The top 1 per cent receive around 30 per cent of the nation's income. The top 0.01 per cent (that is those in the top 10,000th of the economy) receive a full 5 per cent.[28] Wealth inequality is even worse. Around 83 per cent of the country's wealth is owned by the top fifth. Parts of the modern economic system have been described as 'feudal', especially land ownership, where 1 per cent of farmers control almost a fifth of the land.[29] From the violent suppression of efforts to redistribute land in the 1970s, through to modern-day precarious contract farming, economic and ecological risk have been forced down the supply chain onto the worker. When floods hit farms and tenants flee, the owners, astonishingly, can still demand rent.

Pakistan is deeply fractured. Fractured by religious sects. Fractured by economic inequality. Fractured along lines of gender, power, ownership and land tenure. It is also, though, still fractured by the legacy of British colonialism. Pakistan was part of British India until 1947. When the British bitterly granted independence following mass nonviolent and violent resistance (including assassinations and the bombing of British administrative buildings), a British lawyer who had never been to the region hastily drew the borders between Hindu India and Muslim Pakistan. In five weeks he said he was done, and 15 million people were left on the 'wrong side' of the border, leading to one of the largest mass migrations in history. Two million people were killed.

Only a few years before, Winston Churchill had ordered food exports out of British India during World War Two, despite warnings that this would lead to a humanitarian disaster. The resulting Bengal Famine saw around 3 million people starve to death. Churchill blamed it on the Indians 'breeding like rabbits'. Over the course of around 200 years of British rule, life expectancy in India and Pakistan dropped. At the beginning of Empire, the area was responsible for a quarter of global trade. The British actively de-industrialised, destroying infrastructure and smashing technologies. By independence, the area accounted for less than 2 per cent of global trade and was heavily reliant on external markets.[30]

Pakistan remains a nation riven by international forces, from the conditionality of international bank loans, to the megatons of storm water supercharged by the global combustion of fossil fuels. But it is also a nation that can draw strength from its shared spirituality and diverse cultures. Pakistan's landscape and history are littered with alternative models of decision-making, and deep, idiosyncratic relationships with ecology. Just as we saw that the floods have increased the reach of mental-health infrastructure and the development of new systems, perhaps climate change and all that it brings could be the catalyst that Pakistan needs to move towards a more diverse, communal socio-economic system. We need new social formations so communities across the nation can improve their mental health and become far more resilient to climate change at the same time. If Green Squad's co-operatives can demonstrate that, in one of the most fractious regions in the country, whilst also promoting gender equality, economic independence and peace-building, there is little reason why replicating it across the nation should not be within reach.

New economic systems are often talked about as a mosaic. The era we are entering, and may have already entered, is going to take us all by surprise. It will be one of natural disasters, food shocks, mass migrations, armed conflict and power struggles. No matter what people tell us, we are not ready for this. It's impossible to be ready. We will have to experiment. One of the best ways to do that is to bake diversity into any new systems we create. Like any complex ecosystem, it is diversity that leads to resilience. If one connection breaks, if one species dies, if one food source evaporates, then diversity means that there are plenty more to call on. Cutting-edge research into mental-health support suggests the same is true for many human minds. Those most likely to recover from mental-health difficulties tend to have dense, diverse webs of support – whether that is through therapy, family, friendships, work, community, pharmacology or more. The unprecedented uncertainty of climate chaos demands this kind of diverse experimentation. Pakistan is a nation of great cultural and ethnic heterogeneity. This diversity and the different relationships each of its fifteen major ethnic groups and their many, many subsets have with the land and each other are huge repositories of knowledge and practice. If we are aiming for a more contented, responsive and agile resilience – for climate and for mental health – what better foundation could we hope for?

6

Mexico: the small universe which is the body

'Our memory also looks for what is to come. It signals times and places. If there exists no geographic location for that tomorrow, we start gathering twigs, stones, strips of clothing and meat, bones and clay, and we begin constructing an islet or, better yet, a rowboat planted in the middle of tomorrow, the place where one can still just barely see the storm looming ahead.'

'And if there is no hour, day, week, month or year on the calendar that we recognize, well we begin to gather the fractions of seconds, barely minutes, and filter them through the cracks that we open in the wall of history.'

'And if there's no crack, well, we'll make it by scratching, biting, kicking, hitting with our hands and head, with our entire body until we manage to create in history the wound that we are . . .'

'[We] are not fazed by the wall's supposed omnipotence and eternity. [We] know that both are false. But right now, the important thing is the crack, that it not close, that it expand[s]. Because the Zapatista also knows what exists on the other side of the wall. If you were to ask them, they would respond, "*nothing*", but smiling as if to say, "*everything*".'

Sup Galeano, *The Wall and the Crack: First Note on the Zapatista Method*, 2015[1]

THE WORD MEXICO HAS branching roots. The most widely accepted origin of the word is that it's an amalgamation of three Nahuatl words, the language of the Nahua peoples that came from Aztlan and were known as Mexica. These three words are *metztli* (moon), *xictli* (belly button/centre), and *-co* denoting a position or place. Some say this refers to the physical and symbolic position of the ancient Aztec metropolis of Mexico-Tenochtitlan, the centre of their empire at the time of European colonization. Tenochtitlan was situated on the surface of Lake Texcoco, meaning that when the moon rose and reflected off the water, the people of the city were suspended in its image.

Lake Texcoco was drained by settlers and its basin is now the site of much of Mexico City, Tenochtitlan having largely been burned to the ground by Hernán Cortés following a brutal ninety-three-day siege in 1521. The battle started when Cortés massacred a group of unarmed dancers, including the dismemberment of a ceremonial drummer, followed later by the kidnapping and murder of the Mexica ruler, King Moctezuma II.

Others believe 'Mexico' comes from *Mexitli*. This is one name for the god Huitzilopochtli, the Aztec god of sun and war. Mexitli's name is in turn linked to the plant maguey (agave), meaning that the name 'Mexico' could translate as *in the belly button of the maguey,* a sacred plant used in ritual ceremonies. Maguey sap, known as 'honey water', is the key ingredient of pulque, an alcoholic drink. In Mexica culture, drinking pulque was considered a way to move closer to the gods, but getting too close to the sacred was understood to be dangerous, so the consumption of pulque was strictly regulated. Like the agave from which its name may well have sprung, Mexico, then, is rooted. Its tendrils spread deep, fanned and searching throughout its patterned soils, sands, rock and magma.

The Mesoamerican peoples had a nuanced and complex understanding of what we might refer to as mental health. They saw that there were both internal and external influences that could lead to mental disarray – some of these were physiological, some were cultural, some were environmental and some atmospheric, whilst others were determined by spirits. As the scholar of indigeneity Sylvia Marcos writes in her book *Taken From The Lips*:

> According to traditional medicine, the body is porous, permeable, and open to the great cosmic currents. It is not a package of blood, viscera, and bones enclosed in a sack of skin like the one which the modern individual 'has'. Nor can the body be the inert terrain of modern anatomical charts. What must be read in the body are signs of relationship with the universe. Inversely, the external world is rich in signals which bespeak the small universe which is the body. Diagnoses are frequently based on the observation of entities penetrating the body or, inversely, leaving it.

Divination, sorcery and shamanic ritual are all important parts of Mesoamerican medicine. So too are hallucinogens. Mexico has one of the world's richest documented histories of indigenous uses of hallucinogens, including as psychological remedies. This is a psychiatric route only now being explored by physicians in the Western world a full 500 years after colonisation. Long before 'contact', as the Spanish colonial invasion is often euphemistically termed, the Mesoamerican peoples were also pioneers of talking therapy. This was sometimes explicitly ritualistic. At other times it was secular and medicinal, an opportunity to talk through feelings for treatment purposes. It could also be extended to include an individual's family and their community, acknowledging to a degree the embedded nature and inseparability of human health.[2]

Nahuatl is still one of the most widespread indigenous languages in Latin America. There are 1.5 million surviving speakers, most of whom live in Central Mexico. Prior to conquest, what is now Mexico was home to a constellation of ever-evolving cultures and a rich diversity of social groups, including the Maya, the Purépecha and the Tlapenec. Before this, the Toltec and Olmec had all risen and fallen. Mexico has been peopled for 40,000 years. Before Cortés' arrival, there were estimated to be around 21 million people in Mexico, around three times the contemporaneous population of Spain. A century later, the number of people alive in Mexico was less than 750,000. Prior to the arrival of the Spanish, the city of Mexico-Tenochtitlan was home to around 300,000 people on its own, far larger than any European city of the time.

Today, Mexico City, still puckered with remnants of Tenochtitlan's temples, is the biggest metropolitan area in the Western hemisphere. Some 130 million people live in Mexico. It is officially classified as a middle–high-income country, but it is one of the most unequal countries on earth, with around half of the country's wealth owned by the richest 1 per cent.[3] Mexico is one of the largest oil producers in the world. The nation's communities and landscapes are scarred by centuries of resource extractivism and export-led growth, from ranching and mining to the clear-cutting of forests, the redirection of water sources and the fallout from industrial pollution.

Indigenous peoples are still one fifth of today's Mexican population, spread across sixty-eight different groups, and they have borne the brunt of this. Indigenous people are over-represented in the prison system, have less access to the means of subsistence, and are more threatened by the effects of climate catastrophe. Indigenous communities have been systematically stripped of their rights and territories, their access to communal lands have been eroded, especially since the implementation of the North American Free Trade Agreement (NAFTA). From Coahuila to Chiapas, indigenous people are routinely sidelined for the sake of industrial profits.

Indigenous people, in Mexico and around the world, are massively over-represented in ecological struggles. As well as being more active on the front lines of the environmental movement, indigenous earth and water defenders are more likely than other activists to be abjectly criminalised for their activities, the targets of police violence and victims of assassination.[4]

The Cancun COP of 2010, much like Copenhagen the year before, was mostly a bust. My team and I had been given access to the talks by 350.org and the government of Kiribati, a Small Island Developing State at severe risk of sea-level rise, whose president had tasked us with helping to make the summit increase the negotiating capacity of vulnerable countries with few delegates. Often, six critical meetings ran concurrently at the UN, but we found that 15 per cent of delegations were made up of five individuals or fewer. Unsurprisingly, these were mostly delegations from nations with a high level of vulnerability to climate change, whose governments often could not afford to send large delegations, let alone match those of the fossil-fuel lobbies. The EU, for instance, had 6.3 delegates for every 100,000 citizens. The continent of Africa had less than half that.

Taming the wolf
By this time, I was hallucinating fairly regularly. The wolf would sit patiently by my side whilst I made transcriptions in the great white hangars, the contents of square brackets being debated over the speaker systems as attempts to hammer out deals went in eddying circles. It gave me an odd sense of exhilaration, tinged with fear, to have this wild thing at my heel, visible only to me, its body tensed and expectant. The wolf was a threatening manifestation of nature's fury, I thought. That felt appropriate in the huge vacuum of a conference centre I found myself in. We were discussing how to avert Armageddon, and yet everything was sanitised. Defanged. The delegates who really cared were mostly muted and dejected in the halls, or else were frantically organising in the corridors between sessions. Lobbyists and richer nations treated the proceedings like a trade fair.

My psychosis increased. I had one vision of a mythical creature at the very edge of the universe, on the boundary of reality and void. I was flung out to the very limits of material existence, and found a beautiful and terrifying being. Beyond it there was nothing. Negation. The abyss. This thing in front of me looked like a god. It had huge, forked antlers and the head of a silver dog. Its body was that of a leopard, a shimmering gold, spotted with dark rings. The wings of an eagle sprouted from its shoulder blades, the feathers gilded at their edges, whilst its paws were black like a panther's, with long white claws. The chimera was completed by a pair of tails. One was a lion's, tipped like a paint brush. The other tail was a snake, its toothy head twisting out into the blackness. I floated in space, mesmerised. As the creature reached out towards me, one long talon extended, I noticed a thin translucent membrane separating it from our reality. It wanted me to touch it, to join fingertips. I knew I shouldn't, but I wanted it. As we made contact, I found myself outside the known universe, whilst this entity was now inside. I had brought it in. There was a curled smile on its lips and it beat its wings. The creature's ice-blue eyes were fixed on mine as its mouth yawned open, revealing in the depths of its throat a writhing black hole. A shuddering, shaking chaos.

After Cancun I met a shaman called Esperanza at her desert ranch in northern Mexico.

Esperanza told me there was a cure for the apparitions. She said that she had had similar demons when she was my age, that I had been rubbing myself too close to darkness without protection. She said I had been spirit walking, stepping through a rent in the fabric of the universe. I didn't see it as dark. It was beguiling and exhilarating, she admitted, but it was dangerous, and I would go crazy if I wasn't careful. Worse still, I would make the whole world crazy if I kept pulling demented things like that into being, instead of leaving them where they belonged.

She told me to fast for five days. No food. No water. No speaking. No reading. No writing. I was sure the human body could not survive longer than three days without water, but she told me she had done it and to trust her. I did. She told me that the first day I would feel like an idiot. The second day my body would feel like a bag of sand. The third day everything would feel like a waterfall. On the fourth day, the wolf was going to come back, but in a tamed form. The fifth day, she told me, I would be elated, so overjoyed that I would not want to come back. But I had to, she warned. If I didn't, I would die. I started that night. On the way back to my hut, a pack of Esperanza's dogs followed at my heels. Protection.

Day by day, it went exactly as she had said it would. Each day the desert would flash between brittle cold and searing heat as I felt my skin compress around an ever-diminishing husk. I helped build adobe structures, submerged myself in the bright blue encasement of the donkeys' drinking barrel, listened to the whoosh of water pulled up the plastic pipe by their windmill and mulled in the concrete darkness of their storage bunker. On the fourth day the wolf came back as I was outside my hut readying myself for sleep. It was massive. Its thick fur crawled with maggots and patches of rotten flesh. I slipped inside and closed the metal door. The light fell away too suddenly. It wrapped its body around my fragile hut, blocking out all moonlight, and through the window I could see a bloody fluid drooling from its mouth, welts and sores oozing on its skin. Its claws were overgrown in yellowing twists. Its tail was nothing but a series of bones and tattered flesh. It scratched at the hut, took the roof in its teeth, and shook. I spent the night on the bed with my knees pulled up to my chest, watching as the wolf circled and returned. Every time it growled, the noise filled the landscape. Coyotes echoed in reply. Trembling, I thought I was going to die.

The next morning, I watched the wolf lope off in the chalky twilight. I left the hut, sat, and drew figures in the sand whilst watching a swarm of tiny bees flitting in and out of a rosemary bush's purple florets. Next to my hand was a peyote button, a type of hallucinogenic cactus the Huichol people have used for thousands of years as a dreaming plant. The Huichol believe the peyote grow in the footprints of the sacred deer, cacti blooming in its loping wake. When the Spanish invaders colonised the area, some Huichol bands took peyote and disappeared into the desert. It staves off hunger, bringing a placid hallucinatory death. Better that than enslavement by the conquistadores. The Spanish called peyote 'the devil's root' because they saw Satan when they took it. The Huichol say you see yourself. When I stood up from the sand, Esperanza was right again. I felt complete. The sun rising over the distant hills gave the rocky reliefs the appearance of robed humans bowing into the valley pass. I didn't want to come back.

At the ranch, I felt blended into nature. Rock, water, wolf. Esperanza offered me water to drink. She told me I would die if I did not drink, but that idea had no meaning for me. All I saw was the new configuration of atoms I had become. She grabbed my hair, pulled my head back and forced water down my throat, like a flash flood. I was securely, horrifyingly, back in the materiality of my body. I thanked her, the first words I had spoken in five days. She shrugged, sighed and smiled.

Drug wars and mental health

Esperanza's ranch lay between two of the cities most ravaged by the drug wars. I had been told that some areas around Saltillo and Monterrey were almost devoid of men because the cartels would come knocking on doors, demanding foot soldiers. If the men refused, their families might be killed. So, to avoid danger, the men had simply escaped, often leaving their families behind. On one drive I passed under a bridge near Monterrey and saw two bodies hanging from ropes around their necks, bin liners stuffed hurriedly over their heads. Military police patrolled the area in trucks with machine-gun turrets welded on the back, cloaked in balaclavas and black helmets. As we approached the ranch for the first time, hurtling down a motorway, the driver swerved to miss a tree lying across the road. 'The cartels do that,' he told me after we had driven on. 'If we'd hit it or stopped, someone would have jumped out from behind to rob us.'

There is a bi-directional link between climate change and drug cartels in Latin America. By some estimations, cartels are directly responsible for 30 per cent of all forest destruction in some countries,[5] particularly since the US military forced drug runners to carve out sections of remote territory for clandestine flights. These runways are often on land stolen from indigenous people and undermine water and soil protection, leaving the displaced communities and nearby conurbations far more vulnerable to climate impacts.

Climate change is also driving more and more people into the ranks of the cartels. The Rarámuri, who are indigenous to

Chihuahua, have been largely stripped of their subsistence life-style in recent years due to climate-induced drought. This has left some with no choice but to take the risky but economically secure jobs offered to them by drug cartels.[6] Those in some cities live in war zones. Family members go missing, gunshots are a constant background noise and everybody is suspended in a state of sustained insecurity and terror.

In the Mexican city of Ciudad Juárez, when compared to the neighbouring US town of El Paso, researchers found a higher frequency and intensity of mental distress for Mexican children for every single condition and disorder measured, on every DSM scale, including anxiety and depression, sleep problems, aggressive behaviour and attention difficulties. In the year studied, El Paso, the US town, reported the number of drug-related killings per 100,000 people as zero. In Juárez, by comparison, there were 2,101 homicides related to organised crime[7].

The mental-health effects of brutal killings, narco-messages (including notes of warning or publicity left next to dead and/or mutilated bodies) and violent interventions by Mexican government forces are severe. It is not only direct victims of violence who suffer higher levels of depression, but society at large. A single confrontation between local authorities and cartels in a municipality can increase depressive symptoms by over 5 per cent across the board.[8] Across Mexico as a whole, depression is the leading cause of the loss of healthy years of life, affecting around 1 in 15.[9]

The murder of Samir Flores

I became friends with a Mexican guy of a similar age to me, Emiliano, during the Cancun COP of 2010. He works with indigenous communities and organisations, focusing on agriculture and ecology. Climate change is a huge concern for him, as it is for the entire country. Mexico is already scorched by heatwaves and desiccated by drought. Its coasts are regularly hit by hurricanes. The nation's diverse ecosystems are already teetering, from the deserts and scrubland to the tropical and subtropical forests and mangrove swamps.

The average temperature in Mexico has increased more than in the rest of the world. It is already 1.5 degrees Celsius above pre-industrial levels. This has had serious ramifications. The last few years have seen repeated episodes of severe drops in rainfall. In 2021, 85 per cent of the land officially experienced drought.[10] Running water became unavailable in some areas, an unprecedented event in several of the affected cities. It led to mass evacuations and one of the worst droughts in Mexico's documented history. The last time there was a drought as bad, famine spread across the northern states. In 2022, as taps ran dry and the cost of bottled water hit the same price per litre as petrol, Heineken and Coca-Cola continued to extract billions of litres of groundwater, mostly for export.[11] In parts of Monterrey, people's taps ran dry for fifty days.[12]

In central-eastern Mexico, in the shadow of the peaks of Popocateptl and Iztaccihuatl, an anthropomorphic volcano-mountain couple named after two mythological lovers who died entwined, there snakes the body of a gas pipeline. The Morelos Integral Project (PIM) is a combined thermonuclear plant, aqueduct and 160-kilometre-long gas pipeline, which started construction in 2012 despite the lack of consent from the indigenous communities through whose land the project was set to run. From day one, there was a powerful resistance. Emiliano told me: 'The pipeline goes under hundreds of towns, feeding two plants connected to an aqueduct that's sucking all the water from the state, all the water that used to be irrigation water for agriculture. It's dispossession.'

Activists started protesting, marching and launched legal campaigns. This stalled the works in places. In others, construction charged on, with and without licences. Some activists, like Samir Flores Soberanes, a friend of Emiliano's, started speaking out and supporting blockades. Samir, a Nahua man, was a land defender and community-radio broadcaster. Despite years of legal and illegal activity by the project's backers to force it through, the thermoelectric plant could not be turned on because of a 200-metre portion of aqueduct whose construction was being blocked by activists, including Samir.[13]

In 2019, in the face of mounting opposition, the government organised a public consultation about the works. Strategically, they neglected to put polling booths in areas where protest had been especially vocal. One such place was Amilcingo, Samir's home village, under which the gas pipeline was meant to run. At 5:00am, three days before the vote, two cars parked outside Samir's home. As related by the People's Front for Defence of Land and Water, with whom Samir had worked extensively, men in the cars called him out of his house. He walked towards them and was shot in the head. Samir Flores died in hospital. He had received death threats for his opposition to the Morelos project. Allies quickly demanded the project be halted, holding the government and the Spanish, Italian and French companies behind construction accountable.[14]

The public consultation went ahead as planned, despite some activist groups setting fire to polling booths. The official government tally put support for the project at 60 per cent. Activists refused to acknowledge the result, steeped as it was in corruption, quite apart from the fact that most of those eligible to vote were not directly affected. A legal challenge put a stop to the pipeline going through Amilcingo and effectively suspended operations,[15] but in 2020 the National Guard forcibly evicted an encampment of water and land defenders based on the Cuautla river, in order to make way for construction.[16]

Samir Flores's murder continues to spark national outrage. Thousands marched in Mexico City. President Andrés Manuel López Obrador, known as AMLO, having named those opposed to the project 'the radical left, who for me are nothing more than conservatives', officially expressed his sympathy after the assassination but retained his support for the pipeline, suggesting the crime could have been committed to affect his administration. This was brutally and openly hypocritical, given that AMLO had opposed the project during his presidential campaign. He had even compared the completion of the Morelos Integral Project to the building of a nuclear-waste dump in Jerusalem. He had explicitly evoked the spirit of the revolutionary Emiliano Zapata, who called Morelos home and who had iconically resisted land seizures in the early 1900s. Zapata, a national hero, was the inspiration for the Zapatista movement. His grave is a stone's throw from the pipeline. Samir Flores was assassinated one hundred years after Zapata's death. Same state. Same mission. Same cause of death. Today, the building in Mexico City that formerly housed the National Institute for Indigenous Peoples is still occupied by indigenous organisers, guarded by bandana-clad activists and covered in banners with Samir's face and Zapatista iconography.

AMLO has been described as a fossil-fuel nationalist. Despite ostensibly ambitious climate targets, the state has doubled down on fossil-fuel extraction and curbed the construction of renewables, even refusing to allow completed wind and solar projects to operate.[17] Resistance has been fierce, and sometimes successful, as with the forestalling of TransCanada's planned gas pipeline across Yaqui territory.[18] But even this came with serious costs: twenty Yaqui people have gone missing and one person was found murdered on the side of the road, killed with a hammer.[19]

In 2021 alone, fifty-four environmental activists were murdered across Mexico. That perhaps makes it the deadliest country in the world in which to fight on the frontlines of ecological destruction.[20] Worldwide, a shocking 40 per cent of those killed as a result of environmental activism are indigenous.[21] The Mexican state has refused to release any valuable information on Samir Flores's murder. Whether it was a privately hired paramilitary group, directly or indirectly linked to the state, or orchestrated by the funders behind the project, his death was meant to send a message. When I asked Emiliano directly for his thoughts on those responsible, he told me, 'it's the many headed Hydra, the political, business, narco, paramilitary monster, the same monster that's causing climate change'. The thermoelectric plant the aqueduct is meant to feed was completed many years ago. It stands like a hulking white elephant, rusting and too dangerous to turn on.

Indigenous 'madness'

I had originally hoped to talk with Emiliano about the mental-health implications of the Morelos Integral Project, the resistance to it and what had happened to his friend Samir. I had been interested in the ways that indigenous struggles against climate projects were impacting those communities, in whether it was harming them psychologically as well as physically, and how. Was the act of defending their communities helping in any way, psychologically or otherwise? Emiliano thought it was more than just a mental-health issue.

Emiliano told me he knew some people could view the mental-health framing as a 'Global North thing'. Difficult psychological challenges inevitably pop up when people have personal involvement in real movements, he said. Engaging in the conundrums of organising is problematic, not just because of what you are having to deal with in the outside world, but because even when you are working with people you care about there are interpersonal difficulties, losses and reprisals. These things happen even without 'mental-health issues'.

One thing that is often heard is that those who defend their communities and territories by opposing 'progress and development' must be mad. Emiliano said that, from a certain perspective, mental health could be seen as a luxury concept for the rich. The problem lay in the victimisation of the fight, and those involved, as implied by the mental-health framing. As Emiliano told me, the indigenous are already so disregarded and discriminated against that their position is only admitted as the 'viewpoint of the defeated'. They are seen as the 'losers' in a war of conquest and so must deal with the damage.

There is also a long history of equating indigenous people and practices, especially spiritual practices and rituals, with madness. The European medical establishment in general, and psychiatry in particular, often provided initial colonisers with an ideological legitimacy for their actions. It is a small step from demonisation to institutionalisation, from savagery to insanity, wildness and incivility to psychological infantilisation. The colonised have also been routinely incarcerated in psychiatric wards as punishment for resistance. From the 'Natives Only' asylums of British India to the Algerian cases documented in Frantz Fanon's *Wretched of the Earth,* indigenous resistance has routinely been punitively pathologised throughout history by the ruling elite, indiscriminately ignoring the common roots of their legitimate anger in colonial violence and oppression. This labelling of the indigenous and those fighting for their independence as somehow 'crazy' has not gone away. Insanity can be viewed as just another kind of primitivism, both of which are ontological labels meant to objectify the other and make their subjugation a moral act. For the greater good.

Out of those crimes committed in the name of order and progress, as Emiliano put it to me, echoing the words of John Berger, 'one of the worst crimes is that of forcing a people to judge themselves by the criteria of their oppressors, and so to find themselves inferior, helpless, and wanting'. Resistance is rejecting the values of manipulation, the terms of the oppressors and history as they tell it. The worst territorial occupation is that of the spirit, the heart and the mind.[22]

For centuries, indigenous peoples in Mexico have been forced to align themselves with the dominant narratives of modernity and Western civilisation. This has been enacted through dispossession and denying access to land and water, the denial of cultural identity and economic coercion. Industrial civilisation, the many-headed Hydra that killed Zapata, Samir Flores and millions of others, is predicated on the obliteration of self-sufficiency in indigenous communities.[23] A core part of this is the need for a secure sense of one's existence and set of beliefs, having a continuous sense of self. For the Huichol, the Nahua, the Maya and others across Mexico, this depends on a consistent identity, but also a rooting in community, culture, history, ritual and place. Simply telling this story through the lens of mental health, or a medicalised, pathologising and compartmentalising way of talking about the mind, could be yet another contortion of indigenous experience into the conceptual straitjackets of the West.

'It's problematic if you try to atomise mental health,' Emiliano tells me. 'Health is deeply social and that's how it has always been seen in these communities.' Much of this book has been about trying to blur the brittle and often arbitrary lines set out in mainstream discourses about mental health, to understand our psyches as themselves complex ecosystems, inextricably linked to social and ecological realities. So, when talking about indigenous experiences of mental health, we need to take a step back. There have been serious psychological ruptures because of dispossessions, massacres and now, climate change and the fight for a better future. But indigenous people across the world have experienced this kind of persecution for generations and 'not gone mad', Emiliano tells me. I think there is an important question here about what the powerful deem an 'appropriate' response to suffering. Psychological suffering is suffering, whether delineated as a mental-health 'condition' or not, 'legitimate' and 'appropriate' or not. But how those experiences are judged, held and responded to in the world hinge on collective consensus cosmologies that permit or prohibit certain styles of being.

In the mid-20th century, a team of academics, physicians and activists started a movement critical of conventional psychiatry and psychoanalysis. It came to be known as the 'anti-psychiatry' movement. Amongst their ranks were people like R.D. Laing, Michel Foucault and Franco Basaglia, the last of whom pioneered a radical approach to how mental asylums were run in Italy: democratising their institutional form, opening the unit up and eventually abolishing the hospital altogether. One of the precepts of anti-psychiatry was that madness, mental illness and psychiatric disorders were, at the risk of oversimplifying, different states of being and no less valid or meaningful than the psychological experiences of those otherwise deemed sane. The movement often drew on analogous comparisons to indigenous cultures' treatment of different states of mind, from the shamanic utility of psychosis, the apparent evolutionary benefits of depression as a way of stripping the world back to a more realistic objectivity devoid of myth and, conversely, the energetic and infectious drive that mania can have on a community, instilling collective experience with direction, meaning and purpose.

For the anti-psychiatry movement, the persecution of the 'mentally ill' constituted a fascistic form of punishment. Pharmaceuticals were a tool of social control, as were forced incarceration and practices like ECT. Having experienced ECT myself, I can attest to its barbarity. Despite messy evidence in the scientific literature, I still believe that the forced seizures saved my life, but even with the modern benefit of a general anaesthetic, ECT is undeniably traumatic. Years later, my memory is still like a map riddled with large spots of darkened obscurity.

I admire much of the scholarship behind the anti-psychiatry movement. I agree with one of their primary contentions, that civilisation itself is the source of much madness, not just definitionally (as in the concept of madness being a fabrication of the civilised), but also in that the way of life our civilisation forces us into generates mental distress. I am also grateful for the political exercise of reframing mental illness, freeing it from the tight shackles of pathology.

But the anti-psychiatry movement introduced its own dangerous concepts and practices. Some of the most vulnerable patients in Basaglia's asylum left in the mornings and never came back. One patient killed his wife in the hospital. I personally know of someone who grew up around R.D. Laing and was given LSD as a child, which was seen by the adults as a liberatory act. This person spent much of their young adult life reconstituting themselves in the wake of having been experimented on. Some feel that the effort to take mental health out of 'the institution' was unsafe beta-testing on already fragile individuals. They claim that rather than freeing people, the denial of access to conventional treatment cost lives. Some may see psychiatry as a pseudoscience, and at times I am inclined to agree, but for many it is an imperfect yet crucial lifeline in a world apparently devoid of alternatives.

Radical alternatives

In 1978, Félix Guattari, a prominent critic of psychiatric doctrine, visited the city of Monterrey for a conference, the city outside which I saw those hanging bodies on the bridge some thirty years later, and which was hit by unprecedented drought a decade after that. The 1978 conference was on alternatives to psychiatry. The attendees were searching for answers about how best to intervene in such a complex system as the mind, given its constant assault by extractivist capitalism and neocolonial subjugation.

Guattari is probably most famous for his work with Gilles Deleuze, particularly their seminal book *A Thousand Plateaus: Capitalism and Schizophrenia.* This was a dense and weaving philosophical treatise that put forward the concept of the 'rhizome': a dense interconnected fabric of being, with no beginning or end, only a middle, a way of understanding reality and interrelation that runs counter to conventional understandings of reality as hierarchical, linear and based in binaries. Guattari, himself a radical psychoanalyst and environmental activist, was singularly important for his work on the ecology of the mind. In his book *The Three Ecologies,* Guattari proposed a new psychoanalytic model, one based on three distinct ecological realms. The first was the human mind, itself a dense and messy web of an ecosystem. The second was the social realm, which included the political and the cultural, the mechanisms of power through which we live our lives. The third was the global environment, the physical reality we align ourselves with and fit ourselves into. This is a gross oversimplification, but Guattari's key contribution here was to highlight the permeable nature of each of these three ecologies, their inherent interrelation and the implications for revolution. He called this 'ecosophy'.

Guattari was actively involved in political struggle. He saw patients, too. He made it clear that his patients were his allies, and that, similar to Paulo Freire's ideas in *Pedagogy of the Oppressed,* it was only by respecting the expertise of lived experience that real struggle could thrive. He knew that further down the social ladder one would regularly find potent manifestations of psychological distress, but that fundamental problems with civilisation could be traced from within those very states, as well as possible emancipatory escapes. For Guattari, it was never a question of going back to how things were, but instead a question of simultaneously alleviating suffering and tracing a path to a different, more fulfilling existence. Today, the most ambitious and tenacious ecological activists are cast in the same mould. They do not simply want to reduce the levels of carbon dioxide in the atmosphere and return to business as usual. They want to respect this disequilibrium for what it is: a powerful signal of the disequilibrium of civilisation and modernity, a planetary demand for us to re-evaluate our collective cultures. A warning we efface at our peril.

There is no one answer to this question, just as there is no single valid way to discuss mental health. As Guattari writes: 'It seems to me essential to organize new micropolitical and microsocial practices, new solidarities, a new gentleness, together with new aesthetic and new analytic practices regarding the formation of the unconscious. It appears to me that this is the only possible way to get social and political practices back on their feet, working for humanity and not simply for a permanent re-equilibration of the capitalist semiotic Universe.'[24]

When I posed the question at the centre of the 1978 Monterrey conference to Emiliano, how best to intervene in such a complex system, he laughed. How do we do it now? In all kinds of different ways. 'We need a change of perspective,' he said. 'We need to resignify and re-enchant the world, because what's happening to us is really sick. But there's a danger in discursive practice because you can easily say – "we're done, because discourse is my practice"'.

'For us,' he continued, 'you have to get on with the work. Everyone feels like the world is ending. You can either feel fine about it, *or* you can get together and set up your own research and development projects and present indigenous technologies as alternatives to neoliberalism. We already have so many of the answers.' This is a perfect example of what Guattari meant by 'heterogenesis': a many-fingered exploration of the universe by actively building alternative internal and external realities. It is also what is meant by the Zapatista demand for 'a world in which many worlds can fit'. The Zapatistas, a group of indigenous communities who formed an autonomous region in Chiapas in the 1990s that survives to this day, have long been an international symbol of alternative political possibilities.

There has been a systematic effort to relegate indigeneity to superstition, but through the collaboration of communities, scientists and activists, the knowledge and expertise embedded in land-based cultures across the nation are experiencing a revival. Before the arrival of Cortés, the entire central valley of Mexico was a managed water system, replete with organic dams and drainage systems. 'Shamans who were able to call the rain,' Emiliano tells me, 'were real water scientists. They feel and know how and where the water is. The scientifically inexplicable bit is really not that big and it is empirical. There's experience that comes from living with the land that is incalculable and unrivalled. How you relate to water is how you understand it – you cannot put your brain and heart away. You know the water flows down the mountain, and that's not a metaphor.' He stops. 'That's another reason why you can't separate out the "mental". In some indigenous languages there isn't even a subject–object separation. When you talk about something you have to talk about it *in relation*.'

Emiliano refers to Carlos Lenkersdorf and his research on a form of social science amongst a Mayan people, the Tojolabales, that is different from its Western counterpart. Lenkersdorf came to understand by some thirty years of living with the local, indigenous people that their language and social science, as well as their wider culture, is predicated on neither subjectivity nor objectivity, but intersubjectivity, embedded in ecological and social egalitarianism.[25]

Nahuatl is an extremely metaphorical language. Things *are* something else. There is also a different set of linguistic rules for animate and inanimate objects. In classical Nahuatl, only animate objects (and inanimate ones that are metaphorically animate) can be pluralised.[26] Rivers and mountains are animate, put into the same linguistic ranking as humans. The disconnection from and domination of nature was a principal factor in modernity's subjugation of nature. Separating the mental from the rest of the human experience, then, could just be an extrapolation of this disembodying and dehumanising, opening us up to further domination. That is why the mind needs to be rewoven back into the fabric of our wider ecology. And, as with any thriving ecology, the ways in which we do that must be diverse and mutualistic. Heterogeneous.

Small farms of less than 2 hectares account for five of every six farms in the world. They cover only 12 per cent of all agricultural land, but some studies say they produce over a third of the world's supply of food. These technologies are often very old and fundamentally land dependent, woven into local ecosystems rather than transposed from distant lands with no connection to endemic ecology. Family farms, in contrast to industrial farms, whilst less likely to be as historically connected to the soil than smallholders alone, reportedly produce 80 per cent of the world's food. Industrial agriculture is not only a blight on ecosystems worldwide, responsible for mass deforestation, soil erosion and pollution, it is also far less resilient to changing weather patterns and natural disasters. Investing in polyphonic alternatives, rooted in knowledge of the land, is a revolutionary counterpoint.

This needs to be fought for. 'Social struggle is a patent example of not being able to cope with reality,' Emiliano tells me, 'but it's a reaction to not coping. The struggle can actually be the process by which you cope.' These fights for the reimagining of relationships to the land are also fights for reimagining the mind's relationship to culture and ecology. 'That kind of resistance is good for the head and for the heart,' Emiliano says. It builds community. It is meaningful work. It creates togetherness. 'This is where we can talk about the mind in relation. You can have a mental imbalance, but mind is unfathomable and affected by relating to the individual, the physical, the social, the chemical . . . but there's also an important relationship between humans and mystery. Many cultures are more oriented towards relating to mystery, rather than just solving a problem.'

Language and living well

My experiences in the desert were a crash course in relating to mystery. What I experienced there was transcendent. I am tempted to say it was otherworldly, but that somehow invalidates what happened, because it was a different experience of reality rather than a phantasmagorical dislocation from an objective 'real'. Esperanza held me through that. We are not normally taught how to handle experiences like that. Sometimes we don't even have the language anymore. When I had a similar experience in Boston a year later and took myself to see a doctor, I was immediately sectioned. Even though the visualisations were far less dramatic, I was forcibly detained, held down and medicated against my will.

The forcing of my mental-health experiences into the civilisational narratives of modernity was itself a deep trauma. I was pathologised, and the magic and meaning that had meant so much to me was stripped from me and made unclean. I had had a relationship to mystery and the unknown, one that was deeply important, but it was suddenly rendered invalid. A huge part of my being was identified as a sickness and something to be rectified. Before that it had been a largely private, spiritual experience.

It has only been by restructuring my life in the service of nature, that initial impetus that pulled me through the fractured lens of civilisation into madness, that I have been able to piece myself back together, changed. A key part of that has been my network of support, diverse and fluid, a rhizome like any good ecosystem. Guattari's work helped me here. He died by suicide at the age of seventy, the year I was born. After having read much of his writing, discovering this affected me deeply, but it gave his work a transmuted force. Recasting my experiences, removing them from pure pathology and into an ecosystem narrative, has been key for trying to resolve what Emiliano called an 'epistemic rupture'.

The indivisibility of the individual psyche, society and ecology are fundamental. Prior to colonisation, according to the scant records we have, indigenous communities on the North American continent had robust mental health, despite initial contrasting claims by the settlers. Euro-American colonists were fascinated by the Inuit phenomenon of *pibloktoq* (Arctic hysteria) when they made contact from the late 19th century onwards. It was later found, however, that much of the so-called madness was caused by contact with colonisation itself.[27] In 1903, the *American Journal of Insanity* published a thirty-two-year study analysing cases of thousands of residents. It found that the Native American population had the lowest propensity to 'insanity'.[28] Today, solid evidence suggests that speaking indigenous languages in Mexico leads to higher levels of community wellbeing, despite the treatment of indigenous people by society at large.[29]

It was no accident that colonisers spent so much time and effort eradicating indigenous languages, including by banning native languages in Mexico for most of the 20th century. As recently as 2006, a girl was hung upside down at school for speaking Nahuatl in class. Still, however, 68 indigenous languages survive in Mexico (with over 300 variations) and 15 per cent of the population do not speak Spanish. But this vernacular violence also applies to language as a way of understanding ourselves, a means of framing the world.

A common source of inspiration in the decolonial movement is the concept of Buen Vivir – 'good living' or 'living well' – and for good reason. It is an indigenous concept, *sumak kawsay* in Quechua, *lekil kuxlejal* in Maya Tsotsil-Tseltal, Buen Vivir in Spanish. The Quechua are an indigenous group from the Andes. The concept is a manual for living that is all about centring life around community wellbeing, ecology and cultural dynamism. Buen Vivir is cemented in the Bolivian constitution, prioritising ecology and wellbeing over economic extractivism, perhaps the most established, successful alternative methodology to a global economy centred on GDP growth. The Ecuadorian government, reflecting on the concept, recently said that 'only by imagining other worlds will this one be changed'.[30]

Buen Vivir is an entire framework, an off-ramp for neoliberal, neocolonial extractivist capitalism. But it is also, more importantly, a different way of being, one that has been coaxed back to the forefront of culture from embedded indigenous roots. Buen Vivir is just one of a plethora of alternative visions of wellbeing and philosophies of life that indigenous cultures offer. 'We have a primitive disposition to wellbeing, as they say,' Emiliano tells me. 'We know how to be well, as humans. The natural human would be that: the being who knows how to live properly.'

Most of us live in a neoliberal cage. But when we spin out, we see glimpses of other realities. These realities can be terrifying. They can be euphoric. They can be uncanny, disturbing, surreal, disruptive, joyous and debilitating. But what they always do, unfailingly, is demonstrate that other worlds exist. If we crush that down and force ourselves to fit back into the world we came from, the very one that caused our spinning out, we will only temporarily subdue disorder. If instead we learn from the experience, if we respect it, protect ourselves and step out into the world anew, then we have a chance to demonstrate through our very existence that this world is indeed one in which many worlds can fit.

PART THREE

7

Regenerative rebellion: climate action as recovery

'Ecological healing really begins in relationship with each other, and in thinking about community, self-determination and creating openness, to discover who we can be beyond the definitions that have been given to us by others . . . We have been de-democratised and need to practice that democratic participation with each other . . . We often engage in some form of activism or organising because we are seeking that healing.'

Guppi Bola, *Planetary Dysregulation,*
Capitalism & Healthcare, 2023

TAKING COLLECTIVE ACTION IN the outside world has documented and incredible power for working through psychological issues. This is especially true when it is bolstered by reaching out for support and building networks of meaningful relationships. This empowers us and gives us genuine agency, both of which are hard to find in an age when dependency and precarity are sold to us via the medium of calculated lies about freedom and material affluence.

There is a spectacular and fortuitous set of feedback loops that turn engaged action into a powerful therapeutic method. This is not a utilitarian transaction, where helping others or doing good just makes us feel better. It is a reciprocal arrangement, like the management of a commons, and it is about finding purpose and meaning by joining communities of care. It is also about creativity, the opportunity to align our talents with our ethics and creating enclaves in which we can more fully explore who it is we want to be. Together, we try to build the relationships we have always wanted, out of new practices and priorities. This leads to meaningful, impactful action: it is fertile territory for the birth of new worlds.

Trauma, physical and psychological, is at the root of much of our mental disarray. Even the most effective trauma therapies known to modern medicine all attempt to close the loop: confronting and resolving, then methodically working through challenges to proactively restructure one's life. This is the basic tenet of many talking therapies, cognitive behavioural therapy, EMDR, sonic reset therapy and recovery coaching. Many indigenous medicinal cosmologies have practised this for generations. Climate change, and all the despair and trauma it visits upon us, is impossible to confront in its entirety. It is a total planetary disequilibrium. It is pervasive, mysterious and refracting. But against the backdrop of a burning world, taking charge of opportunities to tackle the crisis hands-on is a powerful way of meaningfully addressing its psychological impacts. Facing up to what is happening helps us to retain and reproduce collective resilience in the midst of chaos, building platforms in shared realities that are healthier than the dismal, frenetic and constrained imaginaries of the present.

This kind of systemic visioning and reworking requires a touch of madness: lunacy of the best kind. In 1913, Theodore Roosevelt wrote: 'There is apt to be a lunatic fringe amongst the votaries of any forward movement.' He obviously meant it derisively but it is time for us to take the idea as a compliment and utilise the power and creativity of different ways of being in the fight for a better world.

The prospect of dealing with the combined climate and mental-health crises is daunting and immense. But the resulting fear is something we can digest, helping us to find the joy on the other side of action. The fear itself can motivate us when it is channelled into a passionate anger, navigated by love. Post-traumatic growth can happen on a personal level, and on a collective level. Relationship-building and collective effort will help us form new elements of our identities, more densely woven communities, along with enriching, exciting and unexpected experiences. These will act as dependable mental flood barriers against the rising tide. They will only become more useful, even necessary, as the state of the crisis intensifies, threatening to tear apart whatever bonds we already have. Disconnection and domination got us here. Connection and equity will set us free.

The people I have talked to for this book have had diverse, but for the most part mutually reinforcing, views about the socio-psychological effects of activism. Some initially struggled to pinpoint and communicate helpful psychological benefits to the work they do. I personally think this emerges, at least in part, from a progressive culture of duty and sacrifice comingling with the lived experience of disempowerment under oppression. When I asked Professor Noam Chomsky whether he felt there were positive mental-health outcomes from activism, for the activist themselves, he said that his own experience of addressing injustice 'means living in a constant state of simmering fury for almost all my life back to childhood, a background that is always there.' Later, he told me that he thought building healthier, regenerative models for movements was nonetheless 'important'. 'You're right,' he said, 'groups of mutual support make a great difference to many people in hard times.'

Kate Raworth, author of the international bestseller *Doughnut Economics* and founder of the Doughnut Economics Action Lab, shared a lot with me. She has struggled with depression and overwhelm in the face of the world's challenges, especially when she was halfway through writing her book. It was 'really hard and challenging to see a possible way through,' she tells me. 'For several critical months I got very low.' Today, Kate's work gives her a 'strong sense of being useful and contributing to the community, empowered by seeing other people so energised and mobilised.' She carries on despite self-doubt and questioning, she says, because everyone is continually bolstering each other with energy, support and care.

I came away from every conversation I had whilst researching this book feeling more grounded, hopeful about the future and forgiving of myself. I am grateful for that, as well as for everything people shared about their own lives and work. A better life needs to be an intrinsic part of the fight. Ideologically, I can't relegate it to the end result of an otherwise thankless struggle. The good news is that it doesn't have to be.

A healing kind of activism

Rhiannon Colvin works at Transition Together. Her work is focused on supporting groups of people to take practical action in their local area to build more sustainable, socially just and healthy communities. Her priority is making sure they feel like they belong, that they are supported and can create a meaningful impact. The way she approaches her work now is markedly different from 10 years ago, a lesson that she had to learn the hard way. She is now thirty-three, but when she was twenty she founded an organisation called Altgen, a ground-breaking endeavour dedicated to helping young people set up their own co-operatives as a collaborative and empowering solution to youth unemployment. Altgen burst onto the scene in the wake of the financial crisis. UK unemployment was at a seventeen-year high, and youth were bearing the brunt of the burden. Having unsuccessfully applied for a host of unpaid internships, Rhiannon was acutely aware of the lived experience of the crisis and wanted to create a positive solution to it.

Rhiannon dedicated most of herself to this kind of work. In the run-up to and during Altgen's initial flurry of activity, surrounded by media attention and the complexities of expanding a fledgling organisation, she was fortunate to be part of a group of friends doing similar work. They bonded over a shared passion for politics, parties and trying to build the alternative future they wanted to see, and for a time this brought both happiness and purpose. She told me:

We had such joyous experiences, feelings of empowerment that come from that sense of doing things collectively. Feeling your agency, your capacity, how you can make things change . . . working and collaborating with people towards a common goal. That was joy.

You're not just talking or going out together, it's not transactional. The collective co-creation and the feeling of collaboration and shared purpose has a different quality to it than what you usually get from normal friendships. Maybe you do stuff together, but do you ever create something?

However, soon activism became her 'only purpose in life'. Anything she did for herself, anything not explicitly for 'the cause' filled her with a sense of guilt and shame. 'I remember these thought spirals about having to do more, but being overwhelmed because you can never do enough. It was insane for my nervous system. It took over my whole being.' At a protest at the ruling Conservative Party headquarters in 2010, Rhiannon found herself side-by-side with hundreds of people. Everyone was kicking at a huge plate-glass window. The pane smashed, splintering into sharp wedges that fell with a glittering crash. A huge cheer erupted. It filled her with a cathartic elation, energy and power. 'Rage and anxiety can be powerful motivators,' she told me. They made her feel 'anything was possible'. But eventually, after the dancing and celebration inside the newly occupied space had dissipated, there was nowhere for the energy to go. The rage turned inwards.

Rhiannon's mentor warned her that she was going to burn out. Like many, she didn't listen. In fact, she laughed at the idea. 'I thought I was invincible,' she told me. 'I didn't have limits.' In my own early activism, before my bipolar diagnosis, I experienced the same thing. People often told me to slow down. They told me to look after myself and be more mindful. I remember thinking, patronisingly, that they didn't understand either the urgency of the issues I was trying to tackle, or my apparently boundless ability to persevere. (Either that, or I thought they were jealous and trying to hold me back.) Moderation would have felt like a cop-out.

Eventually, the unsustainability of Rhiannon's constant action and limited support caught up with her. 'I crashed and burned,' she said. 'I couldn't get out of bed. It was such a strong physical rebellion of my body.' She had to drop all of her commitments, move back in with her parents and for two years she focused almost entirely on recovery. Initially, she found the medical help offered to be borderline useless. She tells me her therapy treated 'depression and anxiety in this isolated, personal way. But I knew there was a process, a reason this happened to me. Part of it was personal, yes, but part of it was also the collective situation we're living in and the trauma that stems from that. I had to work two other part-time jobs whilst running AltGen just to pay my rent in London. It's no wonder I was exhausted.'

She eventually managed to crowd-fund money to do a Master's degree. Tragically, the institution's problematic culture did not fit her needs, and that made it difficult to feel safe despite its alternative teaching.[1] She had to leave, but decided to spend a year of self-study completing a Mistress (not Master's).[2] Her topic? The social determinants of ill-health. Rhiannon's dissertation presented a systems map that has since reminded me of a later (unconnected) diagram published in *Nature: Climate Change*, also about the systemic drivers of mental health.[3]

Since her burnout in 2016 Rhiannon has lived with the chronic illness Chronic Fatigue Syndrome/ME and this has necessitated that she build a different (or healthier) relationship to her work and activism: a healing kind of activism. Rhiannon now works for Transition Network, specifically its new UK wing Transition Together. She tells me: 'I've realised that my purpose is to create change, but it is also to experience the simple joy and pleasure of being alive and to share that with the people I care about. It's not about having to choose between myself or the cause anymore, it's about making space for both and about doing work in a way that brings me connection and joy.' Then she smiled and told me: 'It's so nice to be asked about the positive mental-health impacts of activism for a change. I can't imagine any other type of work giving me such meaning and purpose. It is vital

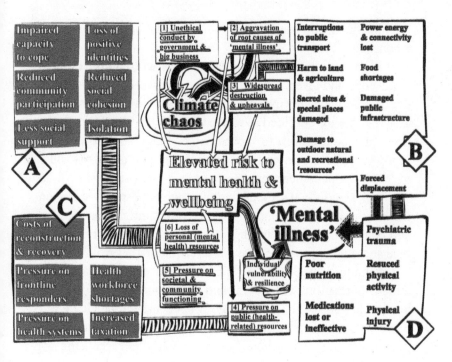

Top-level causal process diagram showing major domains of distal, intermediate and proximate harm linking climate change and mental illness, adapted from HL Berry et al, (2018), *Nature: Climate Change*, Vol 8, 2018, 'The case for systems thinking about climate change and mental health'. The authors' original explanation of the diagram:

'Each domain affects the next and, through various proximate mediators, also independently elevates mental health risk. A: Factors that interact to produce loss of personal resources. B: Factors that interact to produce widespread destruction and upheavals. C: Factors that interact to produce pressure on public (health-related) resources. D: An example of proximate mediators that lie on sub-system pathways connecting particular domains to mental illness. Individuals' underlying vulnerability (or resilience) increases (or decreases) the risk of mental illness as a consequence of climate-change-related exposures.'

that this healthier kind of activism is support by the organisations and movements we work within. Transition has really thought this through. They proactively support us to avoid taking on too much. It's one thing to have personal boundaries when you re-enter this work, but if the organisation doesn't support it, it's a constant fight.'

With Transition Together, Rhiannon has written articles, held workshops and given talks on this, her new work, largely a practice of introducing her insights into movement building. Transition Together has also decided to weave Community Organising practices into its approach to movement building, supporting local groups to have one-on-one conversations with potential new allies to 'meet people where they're at', letting them start gently and not feel excluded. 'It's about finding a personal story,' Rhiannon tells me, 'a reason for wanting to get involved and find connection and common ground. Tapping into people's feelings of injustice, rage and anger – but also love – is a technique that really works.'

Rhiannon is supporting people in the movement to stop initiating conversations with dogma and theory, substituting that with lived experience. The work is no longer just about giving out flyers or setting up talks. It involves going out and talking to people and supporting them, hoping to help them access the joy, connection and empowerment that comes from doing meaningful, practical things together. 'Taking action on the individual level can feel overwhelming and lonely. Looking at the whole global picture can feel disempowering. But if you work with others at the community level it feels manageable and that can create empowerment and active hope.'

Rhiannon insists on including a power analysis and strategic thinking. You have to figure out what issues you are trying to address, and how to actually help people, build resilience and create cultures of care. Modern climate movements cannot just be the reserve of the privileged, people with racial, disability, gender, class or time freedoms. They must be intersectional and accessible and deal with immediate threats, joining explicit links between climate and race, the cost of living crisis and fuel poverty, for instance, with immediate practical actions as well as movement building. As Rhiannon says: 'It's still an open question in my mind, whether we can transform this movement into something that's really able to shift things.'

146

Finding a purpose in the corporate labyrinth

Fiona Callaghan came to a similar, if contextually different, real-isation a long time ago. Fiona is a lawyer who has spent a lot of her career working on development projects for corporate clients in London. She is Catholic and a mother of two. She increasingly questioned the value of her work: 'Whilst the legal challenges were interesting, once they were mastered then for more than a decade I felt I was not contributing in a valuable way to things that mattered to me.' Through frustration, through action, Fiona found a route out. During our conversation, she thanked me for inter-viewing her, then said: 'You should be interviewing someone from . . . I don't know where. I'm really not doing incredible work.'

That couldn't be further from the truth. It is just that she has been doing it all in the shadows. Fiona has always wanted her work to align with her values, but she became professionally stuck. It is a pervasive feeling, but one that is rarely talked about publicly. One survey found that nine out of ten people are willing to earn less money in exchange for doing more meaningful work. Not only do most professions lack an explicit ethical component in line with employees' values, but somewhere between 20 and 50 per cent of people believe their work is 'useless'. A quarter of British jobs 'lack meaning' according to a YouGov poll, and two-thirds of young people are struggling to find what they consider 'meaningful' work. For many, doing good work means being employed by a charity or a campaign group, where pay is typically low and vacancies hard to come by. Otherwise, it means volunteering in one's spare time, and most of us have little of that as it is.

The primary earner in her household, Fiona felt responsible for her children and their future. She often worked sixteen-hour days for corporate clients, year in year out. Climate has been on her mind for a long time. For a decade and a half Fiona has been reading copious amounts of news about climate change. It wasn't just that she cared about the issue. 'I was scared,' she says.

When the Carbon Reduction Commitment Energy Efficiency Scheme came out in 2010, Fiona saw the potential value of applying her legal expertise around buildings to questions about climate. 'That was the first link,' she told me. 'Real-estate emissions are 30 per cent of our emissions and buildings consume 40 per cent or so of energy.' At that time, she says, very few people in the industry were looking seriously at the carbon emissions from buildings or the difference between tearing down existing buildings and starting afresh versus retrofitting. Nor were they thinking about how to find a fair way to help landlords and tenants co-operate on improving the energy efficiency of rented buildings.

She was not a perfect fit for the role, but no one else seemed to be volunteering to look at this issue. Policy makers were not providing an external push, and nor were campaigners. Draft excluders and new models for lease agreements are boring. With some exceptions, the environment team at her law firm could have done something, but they didn't have the property expertise required. 'I just stepped up because I was interested and I understood the property bit,' she told me. 'I was kind of sticking my neck out, but it was a new piece of legislation and nobody knew anything about it.'

'After that I continued to keep up to date on sustainability-related policy and regulation which affected the built environment. I did that outside of chargeable hours, just because I wanted to do it and I wanted to be involved,' she says, 'and at the time our environmental law team were busy with the kind of things they did back then which tended to relate to contamination and pollution rather than climate-related sustainability where there was little regulation and not much call for legal advice. During the 2010s not many property lawyers were thinking about these issues and I was lucky to be asked to sit on the sustainability groups of a couple of industry bodies which meant my knowledge grew and I had a place where I could contribute in this area.'

Working 'full, full, full time' and dedicating evenings and weekends to climate law was not sustainable, though. She felt distanced from and unavailable for her children. She neither knew how to handle the emotional and psychological weight of climate change herself, nor what to do about it practically. One of things she grappled with the more she read about climate change was whether to talk to her children about it. She stayed reactive in conversations with them about it until a couple of years ago when she broached the subject with her then twenty-one-year-old daughter, Niamh, on a boat in the Outer Hebrides. They were chopping over waves between Skye and Lewis. The waters are said to harbour the Blue Men of the Minch, sea spirits that look like men except that they are blue from head to toe, and are perpetually in search of ships to sink and people to drown.

Recounting her thoughts that day, Fiona told me: 'You don't want to say to your children, "things are really serious, it's not clear what your world is going to look like". I could be more gloomy than that, but let's just leave it there.' She continued:

> At the same time, I don't want to pretend it's not happening. I know Niamh is not naïve about these things, so I asked her casually on the boat, whether she and her mates talked about climate change. 'Yeah,' she said. And what was odd was that she was so undramatic about it, so matter of fact. So I asked what they talk about, what they think.
>
> She said: 'I would love to be a mother, but I don't think I'll have children because I don't feel that I can keep them safe. I would adopt, but I wouldn't have my own children.' And I said, 'God, that's a pretty big thing,' and she replied calmly with: 'Yeah, well, that's just how it is.'
>
> Niamh told me she has one friend whose idea of happiness is to have a wife and children, and so he says he's going to do that and give them the best life he can. But he will always keep with him some way of painlessly killing them all if life becomes unfeasible.

As an interviewer I was shocked hearing this. It wasn't because it was novel, but for quite the opposite reason. Others had outlined the exact same strategy for climate Armageddon to me only recently: a family plan for instant painless death should things get too hairy. Fiona continued: 'I thought, my God what have we done? They don't feel they can even have children, or continue the human race. And Niamh wasn't crying. She wasn't dramatic. She said she's got other friends who say, 'well, we're all fucked – so let's just enjoy it'.

When Fiona returned from the Hebrides and went back to work, the day job continued to be demanding, the sense of purpose she wanted was still lacking and she found herself increasingly despairing at the amount of new development that was not meeting the needs of local people and which was adding to the amount of embodied carbon emissions.

She was ready to jump ship but her firm had begun to take steps which recognised the importance of climate change to its clients and its work. It introduced new roles: heads of climate for every department. They asked her to be the climate lead for real estate in addition to her existing role. She knew it would mean going even further above and beyond, but she thought: 'Yeah, I'm going to do this actually.' Then, after a particularly gruelling period of work: 'I decided to give up the day job. I told them I was going to resign, but that if they wanted me to stay on, doing anything, then it was going to be climate and to their credit they agreed.' Disentangling herself from her previous role and finding a new way of working that used the knowledge about climate issues relevant to the built environment that she had acquired in her own time over more than a decade was good for her, and good for the climate. But it was a long journey, often thankless and sacrificial. I asked if she regretted it. 'No,' she said. 'Not at all.'

Now she is part of conversations with leading industry representatives 'who all get it and are pushing for the right regulation and policy' whilst also collaborating and sharing best practice. She is getting into the details and dealing with clients on climate issues face to face, not only keeping them abreast of climate policy but trying to show them the benefits of going beyond it. 'Going out and doing that has made me feel differently about climate change,' she told me. 'I feel like I'm doing something about it. I have some agency and some sphere of influence.'

Being able to support others in the space to do similar work has been empowering, too. 'It's not that hard anymore because most people are on the same page, they just don't know where to start – so you just start telling them. It really has helped. Also I have a bunch of new colleagues doing climate-related work in other areas of law who are dedicated and inspiring, which is wonderful.' When I asked her why she didn't just give up, or go down the hedonism route, she told me:

> Because I'm not without hope. There has got to be hope in there somewhere. I think there's something we can do to make it better for our children. Young people have got such a bum deal. And my generation just didn't have to worry about any of this – none of it. My generation had to worry about how to build a shelter for nuclear fallout, yes, and the pamphlets through the door for that . . . but the nuclear issue felt somehow easier – it just took one person not to press that button. It was fairly black and white and that was easier to deal with, somehow. This feels like the whole planet is in disarray and deterioration. It's a different level of threat.

Importantly, she feels that what she is doing matters deeply for her daughter and her son. For the first time, according to Fiona, her daughter 'might even vaguely say that she's proud'. As a whole, she says her life has become significantly better after struggling through the messy transition towards officialising her work on climate. 'I feel much more positive about myself.'

As I stepped out of her front door that night, under lamplight, I couldn't help but reflect on quite how lonely and atomised Fiona's journey had been. She fought a battle for over a decade to have something recognised, something that people around her did not value much. A lot of activism involves the same relentless process.

But even with the constraints of paying bills and career progression, Fiona managed to persevere. She acknowledges she has certain privileges, but the route can still be far easier for many.

Other barriers to active work on climate change abound, whether in the form of illness, discrimination and structural oppression, but there are ways for us to find a niche and support each other as we carve that out.

The never-ending walk towards utopia

Too often climate action is seen as the preserve of the privileged. We are encultured by a system that breeds these injustices and we have to navigate them within the movement as well as between our work and the outside world. As Rhiannon wrote in a reflective piece about leaving university: 'Unless we proactively put structures and processes in place to create new norms it is unlikely that these norms will change randomly and suddenly.'[4] She has carried this sentiment into her work. This is an important focus and, as later chapters will explore, it is key that we learn and practise building alliances, connections and community into whatever work we do – whether that is changing career tack or physically dismantling fossil-fuel infrastructure under the cover of darkness.

We get to play with different configurations of the world as we figure out what kinds of worlds we want. Eduardo Galeano's never-ending walk towards utopia must be free of the constraints of capitalist imaginaries; a liberated adventure of communal meaning-making. Our culture's dominant ways of knowing exclude those it deems a threat to extractivism. It has done so for centuries, eviscerating cultures, peoples, landscapes and the cosmologies contained therein. Extractivism extracts. Its purpose is to reduce the living world to capital, including human bodies and minds. It is the opposite of healing. It disconnects and dominates whilst claiming to enrich. We find ourselves in a state of mental disarray and global disintegration. This means we often forget the dizzying array of skills and knowledge we have that could be ready to apply to regenerative rebellion.

Both climate despair and climate trauma highlight the irreconcilability of extractivism and a thriving earth. Our disjointed minds are signals of our world's imbalance, experiential proof of our interrelated, diffuse, empathic natures. Our minds are reminiscent of the canaries that coal miners carried down to the depths of the earth to test for poisonous carbon monoxide. Canaries have extra air sacs in their bodies, so they absorb oxygen on both the in breath and the out breath. This means they also absorb anything else that is in the air, including poisonous gases. As sentient sentinels unwillingly exposed to lethal vapours, they would flap their wings against the spindly bars of their cages, their death throes signalling to the miners that they should escape. The bright, receptive, incarcerated birds took their final breaths in darkness, encased in tunnels bored within the earth on behalf of fossil capital. We may be similarly receptive, but we will not die so easily.

How to change a system, then? Systems theory, for a start, tells us much about maintaining dynamic equilibrium in complex systems. It also points to how to tip things into entirely different states. The Land Back movements of today, for instance, are about far more than who owns ancestral lands. They are also about questioning the very concept of ownership, what know-ledge/truth/expertise are, how to relate to the living world, and how to access and distribute the means to be and stay well. These are struggles that are acutely strategic, undermining the dominant system as they bring others about. It is the same for the Zapatistas. It is the same for the people of the Autonomous Administration of North and East Syria (Rojava). The Black Panthers had the same methodology. Far from being just reform movements, they actively reimagine and recreate the whole world, from the micro to the macro. It is only by redesigning and rediscovering altern-ative ways of being, and practising them in the real world, that we can shift into healthier more fulfilling worlds beyond the dank cold metal of the canary cage.

Systems exist at many levels, similar to the visual patterning that shows the structure of the human brain to be remarkably similar to the distribution of galaxies in the known universe.[5] Our physiology depends on dynamic systems, from the homeo-static controls that regulate our body temperature to the psychological feedback loops and adaptive strategies that enable us to cope mentally. Families are systems. Communities are systems. So are societies, political and economic configurations. So are ideologies, ecosystems and the climate. Each nested system impacts on the other, much like the trophic cascades in complex ecosystems. The most famous example of a positive trophic cascade happened after wolves were brought back to Yellowstone National Park in 1995. The inter-species dynamics that resulted led, remarkably, to major improvements in the health of the Yellowstone river. The graphic below is adapted from one developed by neo4j; it visualises some of the dynamics that led to these surprising changes.[6]

Complex natural systems are vastly more intelligent than any single species, no matter how clever we humans think we are. Each nested system, from the dynamics of our biology to the dynamics of the planet, can be shifted or tilted this way and that. This is different from just improving the existing system: we are talking about wholesale systemic change. This means a period of change followed by a new dynamic but relatively stable state. This takes a different kind of seeing. Instead of envisioning an endpoint or just rejecting the status quo, as a lot of traditional

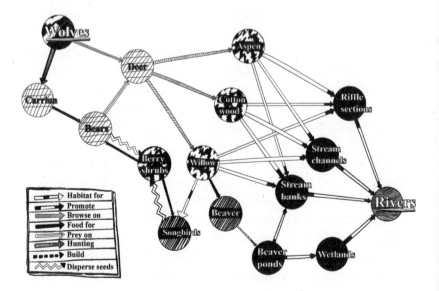

campaigning does (often to great effect), systemic change is about the process of taking action that creates a fundamentally different world as we go. It is about being relational and providing support to everyone involved as a reflection of the principles we want to engender. It is about continually learning from, adding to and adapting to our collective work. Systems work respects complexity and uncertainty, being tactical and action-based but also reflective and humble. It is not about 'winning'. Nor is it about just talking forever. It needs some organisation, as Jo Freeman beautifully illuminates in *The Tyranny of Structurelessness.*[7] Existing in altern-ative social formations, in turn, helps us to hold different perspectives on reality and to create new and better ones.

Understandably, given the enormity of our predicament, huge numbers of people have worked to figure out how to change a system strategically and compassionately. Many of them can be loosely categorised into three parts.[8] For our purposes – organ-ising that builds psychological resilience whilst simultaneously bringing about transformative change – we can define these three parts as follows:

First, we **Resist**. This means stopping harm, whether that is physically dismantling fossil-fuel infrastructure or simply resting and recuperating. Stopping the distress- and trauma-causing behaviour can be therapeutic, if done right.

Second, we **Reconnect**. None of us can do this alone. If we are radically inclusive we can find unity in our diversity. There is a wealth of knowledge and experience amongst those most affected by climate and mental breakdown, and even more in systemically connected struggles where people have been fighting these existential battles for centuries. Connecting builds community and capacity, whilst also learning how to care for each other in practice.

Third, we **Remedy**. This is a process of imagining and imple-menting different worlds. The extractivist system aggressively destroys any potential alternatives that threaten it. But not only is another world possible, it is imminent. We have so many viable pieces of the future lying at our feet. The act of putting them together is radical, an adventure, and it is a creative realisation of a world in which many worlds fit, as the Zapatistas say.

Avoiding turning inwards

I want to briefly highlight what this is not. Namely, self-soothing inward-facing action on its own. We have a tradition of drinking in extractivist values, especially individualism, when we try to change the world. It is quite common to find friends and loved ones who are pressured by ecological devastation and personal guilt, even shame, swayed by the consumer capitalist promise of individual self-improvement. We have an entire behemoth of a system operating on the idea that status, individual responsibility and self-sufficiency (in the capitalist macho sense, not the ecological sense) are aspirational. It is not surprising then that so many of us turn to well-meaning coping strategies like insular spirituality, changing personal consumption habits and occasionally engaging in clicktivism or showing up on the edges of protests. This kind of action can be helpful. But it rarely reproduces itself, inward-facing as it is. It is often sacrificial, can have negative mental-health consequences and very little meaningful impact on the outside world. It can be a vital part of a wider project of repositioning ourselves in the world, living more in line with our values, but sometimes it can end up sedating us, closing off possibilities for connection and post-traumatic growth.

Take reducing our personal carbon footprints. Only about 2 per cent of global emissions come from aviation and 14 per cent from livestock.[9] An understandable but still disproportionate amount of attention has been paid to choosing not to fly and to cutting animal products out of our diets. Even if these sectors emitted more, individual consumer choice isn't a good mechanism for bringing about rapid systemic change. It essentially puts all of its eggs in allowing the market to adapt to consumer trends, one person at a time, rather than forcing the system to fundamentally overhaul a dysfunctional supply chain.

I was recently listening to the radio and Mary Portas, an otherwise stalwart campaigner for alternative economics and localism, made a comment that I found upsetting. She said: 'Every one pound spent is a vote for how we want to live.' Any distillation of social activism to ethical consumerism is depressing to me. If we agree to fight on the enemy's territory, using their logic, we have already lost. Green consumerism is an oxymoron. As a capitulation to market forces it is anti-democratic in the worst sense, giving more power to those with more money. It ignores the in-built logic of income and wealth inequalities, competition and economies of scale that make it nigh-on impossible for truly ethical businesses to thrive under the existing system.

Spending money ethically is good, provided it is not seen as an endpoint in strategic intervention. We should ask whether it detracts from or replaces more outward-facing activities. Ethical consumerism does have the potential to dominate over other focuses, in ways that can harm us as people as well as the wider world. We must guard against turning too far inwards. I say with love that we cannot allow ourselves to become fixated on changing our own consumption choices, whether in terms of diet, fashion, travel or whatever. It can lead to shame, isolation, alienation and fatigue. At its worst, it can lead to disconnection and domination. If ethical purchases and lifestyle choices are healing, that is wonderful, but we have to be careful. We are not the problem, the system is. We need an expansive, exploratory set of ideas and actions that can help us to build real alternatives in the interstices of our existing reality.

About nine years ago I was in a state of deep psychosis. I was on the way to dropping out of university because of mental-health issues for a second time. I was living on my own in a flat in London. It was a dingy space, but better than the squat I had moved there from. The walls were thin, as was the ceiling, and I could often hear the family above me arguing. In the state I was in, I was convinced they were arguing about me. I rarely left the place for the month I lived there, except to walk alone at night through the wintry streets. I would march along the pavements, certain that somebody was following me, until the scrutiny and threat became too much and I would hurry home.

I would live in darkness, never turning on the lights for fear someone was watching, and so would navigate with the torch on my phone. I was very alone, kept company only by a bouquet of lilies my mum had given me on moving in. Slowly they turned from white, to beige, to brown, the anthers shedding their rust-like pollen on the table as the water in the vase went dark and began to smell.

None of this felt relevant to me, and nor did the urine-stained water that started leaking down my walls in tidal ripples from the flat above. In the midst of this all-encompassing madness, one thing felt like an anchor. On my night-time walks I would collect shards of metal, cogs, screws and folds of brass that had fallen off construction vehicles, together with bike chains, coils of bark and bright shards of plastic in primary colours. I would arrange them on my bedroom floor, making little patterns that became shrines. They were islands of calm and order, of quiet beauty and control surrounded by a raging torrent of uncertainty and threat.

Clearly my version of 'self-soothing' at that time was extreme, to say the least. But I feel like there is a danger that some of us end up contending with climate chaos by 'cultivating our own garden'. We get on with our lives whilst all around us the house is falling into ugly disrepair. But rather than do anything structural, anything big, we turn our sights inwards. We create a home world and a personal behaviour that is line with our vision of a better life. It is a survival strategy. But is there a chance that doing so stops us engaging in collective action to bring about the bigger-picture, systemic changes that are so clearly needed? I was only saved from my own personal insanity by family. I didn't have the capacity to escape on my own. They visited one day and saw the state I was living in, the wild-eyed terror I had grown used to, and forced me to move out.

What I worry about is our becoming trapped in a cycle of self-soothing that holds us back from being more ambitious, more radical and more realistic. Importantly, that trap would also mean that we miss all the incredible experiences and relationships that more outward-facing actions can offer us. It would be very convenient for those in power if all those who care about climate change felt satisfied with pouring their concern into individualised action. The same is true for mental-health issues. A more systemic series of interventions are needed than everyone downloading Headspace and taking the right medications. Both of these crises require a protracted effort to build alternatives. I am not talking about self-sacrifice and puritanical idealism here but rather about becoming involved in creating a different future, in whatever way works for you, along with all the benefits that can bring to your own life.

Many of us are time poor. Many are struggling financially. Many are victims of incarceration, abuse, discrimination and oppression. Getting involved in this stuff can seem unrealistic. But just as mental health cannot be codified as a Global North thing, active engagement in organising cannot just be for the most affluent. The most powerful movements throughout history have been driven by the dispossessed. Today, we lack enough meaningful communities built on radical inclusivity. Both community and radical inclusivity are things that can make it possible to involve ourselves sustainably. Communities of care, like mutual-aid groups, can also simultaneously address the immediate precariousness that keeps a lot of us on the outside.

Collective action and movement building generate a surprising amount of joy and a sense of belonging. There is a lot of evidence, from those on the frontlines and in academia, that engaging in social activism can have powerfully positive effects on our minds. As Rebecca Solnit, author of *Hope in the Dark*, remarked in a recent article: 'I have often met people who think the time I have spent around progressive movements was pure dutifulness or dues-paying, when in fact it was a reward in itself: because to find idealism amid indifference and cynicism is *that good*.'

I will be honest here: there are definite dangers. A lot of activist communities have sadly ingested many of the same ideas and practices responsible for driving us to states of crisis, often just through osmosis. They have pushed people to excessive levels of ambition, status-competition and into repeating patterns of oppression and discrimination even whilst they are saying all the right things. Few of the activists I started working with a decade ago are still involved in movements, turning instead to other practices and disciplines because of burnout, trauma and despair. We need to learn to care for ourselves and to care for one another. Thankfully there are lots of examples of people doing just that.

We each need our own visions of utopia if we are going to set off in search of them. But to paraphrase john a. powell: there is no such thing as perfect and, even if there was, each person's version of perfect would be different. What we probably need, then, is an understanding that our progress towards a better world will be a process of radical experimentation and improvisation.

We have no solid blueprint for this, at least not one airtight enough for immediate worldwide implementation. In many ways that is a positive thing. What we do have is a vast collection of ideas, movements and actions that could help us shift course and, as we do so, contribute to building a series of alternative systems that can realign human society with ecology and make us healthier in the process.

Being involved in a movement for tackling climate change, improving mental health and building a better world has to be a positive, meaningful and healing experience. It has to be fun and regenerative. Otherwise, what on earth are we doing, and how on earth are we going to get through this?

8

RESIST: how to stop the bad stuff

'Those who profess to favor freedom and yet deprecate agitation are men who want crops without plowing up the ground; they want rain without thunder and lightning. They want the ocean without the awful roar of its many waters . . . Find out just what any people will quietly submit to and you have found out the exact measure of injustice and wrong which will be imposed upon them'

Frederick Douglass, *If There is No Struggle, There is No Progress,* 1857

Lela: 'I'm sorry but if it's fun in any way it's not environmentalism.'
Environmentalist: 'What about blowing up dams?'
Lela: 'Okay . . . that is fun.'

Futurama, S3 E9, 2001

A. Greening urbanity: tree planting and guerrilla gardening

IT MAY SEEM ODD to start with planting things immediately after a Frederick Douglass quote, but radical greenery is subversive as fuck. Playing with our environments can simultaneously put a spanner in the works of the ecocidal machine and lead to happier, healthier spaces. It also rends open the fabric of our constrained imaginations. There are few images more metaphorically complete than a sapling pushing up through concrete. My dad used to put superglue in the locks of major banks' main entrances, under the cover of darkness. It gave him a thrill, but it also meant he had a different kind of relationship with the city he lived in. The buildings weren't just buildings. The streets weren't just streets. The urban landscape became a playground, one he was an active participant in. His actions could have got him arrested, and we will get on to similarly risky stuff, but let's start by wading in shallower waters.

A friend of mine, Max Jourdan, plants trees in London. He has no permit. He just goes out and does it. 'It's a fun hobby for me,' he says. 'Feels like time and energy well spent in the frenzy of a worklife. And of course there's the pleasure of trans-gression – smashing up paving stones with a sledgehammer and digging up London clay with pick and shovel. Pretending I'm a man from the council in blue overalls and high-viz jacket on an important mission.' It's rebellious, but it's also community action. 'The trees I plant, and must nurture, also connect me intimately with the landscape of a neighbourhood and the people I invari-ably meet. Everyone seems happy to see green growth emerge from grey sidewalks and waste grounds.' He has been doing this for about twenty-five years, mainly trees he grows from seeds or cuttings. At the moment he is into giant Echiums flowers. It's good to experiment, with some guidance, making sure we add life that is viable as well as beautiful. It's important, too, not to block any accessibility for wheelchairs on pavements and to stitch new life into our surroundings in the best way we can. Not far from where I'm writing this, Max planted a Giant Sequoia. He brought it from California and it now stands fifteen years young in the shadow of Trellick Tower. He calls it a 'future millennium giant'. He says: '[Planting trees that]will hopefully outlive me elicits a warm and fuzzy feeling'.

There are tonnes of online guides for how to reverse the tendency for brutal destruction of biodiversity in cities and beyond. One of these is guerrillagardening.org, a blog about illicit tending and planting in public spaces around London, full of tips for how to reclaim urbanity for plants and people.[1] Seed bombs are great. These are spherical cakes of earth, clay, fertil-iser and seed in some combination that can be chucked into derelict sites so that plants can take over again. This idea has been used in Kenya, where SeedBalls Kenya drop thousands of tree seeds encased in clay onto arid and often private land – sometimes from helicopters.[2] Other organisations around the world use seed bombs as a quick and convenient method to begin rewilding bio-scarce landscapes.

Grouse-shooting moors owned by the UK's elite take up around 8 per cent of the entire country's land and are devast-atingly devoid of wildlife.[3] Some people target golf courses, which, in the US, cover more land than New York City, Detroit, Chicago, Los Angeles, San Francisco, Miami, Austin, Boston and Minneapolis put together.[4] TikTok is full of tutorials and demonstrations on guerrilla gardening. As TikTok creator and activist Jeremiah Jones told *Insider* magazine: 'Our health, emotionally, and our physical health depends on natural green space. So [Guerrilla Gardening]'s about amplifying that and making sure that everybody has access.'[5]

The act of doing something loving and rebellious, like kick-starting a mini ecosystem, is powerful. The spaces that result can bring real benefits to communities, especially those normally deprived of green space.[6] Max's first foray into this space was raking gravel and planting corn on an abandoned building site. He tended it and watered it from an abandoned galvanised tank that trapped rainwater. Birds came. In the summer, the ears were nearly shoulder high so he and a friend picnicked and sunbathed. An 'Eden', he says, until JCB diggers levelled the ground and tarmacked over it for, I kid you not, a parking lot. Urban green spaces reduce air pollution, increase climate resilience by regulating temperature and water flows, whilst green access also has a strong impact on cognitive performance, depression, anxiety, cortisol levels and general improvements in wellbeing and emotional resilience.[7] Unilaterally bringing them into existence is relatively easy, fun and empowering. Get on it. Fight the greyness.

B. Tyre extinguishers

Globally, Sports Utility Vehicles (SUVs) are responsible for 900 million tonnes of carbon dioxide. If these massive cars were a country on their own, they would rank sixth in the world for emissions.[8] Alongside planes and coal-fired power plants, SUVs have long been an effective target for climate activists who want to make a significant rather than symbolic dent in atmospheric levels of carbon dioxide. In 2007, a Swedish group started deflating the tyres of SUVs in Stockholm's most affluent neighbourhood.[9] It caused outrage, became the target of a media frenzy and a slew of death threats from SUV owners. Ultimately, though, it helped reduce sales of some brands of SUV in Sweden by more than 25 per cent.[10]

More than a decade later, a similar movement sprang to life in the UK before quickly spreading around the world. Tyre Extinguishers is a decentralised, autonomous movement. 'We have no leaders,' they proclaim, 'anyone can take part, wherever you are, using the instructions on our website.'[11] Those who want to get involved are given clear directions on how to unscrew the air caps on SUV tyres (lefty-loosey, righty-tighty), insert a small stone, bean or lentil (if you want to keep the hippy vibe alive), then re-screw the cap until you hear the quiet hiss of air escaping.

This is usually done at night, preferably in pairs, and explicitly targets the most ostentatious SUVs whilst avoiding any that might be used by disabled people, tradesfolk or minibuses. The website provides a leaflet to print out (available in thirteen languages) and put in the windshield of any immobilised cars, explaining the reasoning behind the actions. 'You'll be angry, but don't take it personally,' the leaflet says. 'It's not you, it's your car.'[12] In less than a year the group deflated the tyres of more than 10,000 SUVs, including an internationally co-ordinated hit on 29 November 2022 that saw 900 cars across the US and Europe 'disarmed.'[13] There are at least one hundred autonomous groups across the world but as one member told me, 'you just print out some flyers, and off you go!'

One participant and organiser told me that it can provide:

A cathartic release . . . The anger and frustration that can build up when you see [SUVs] amassing in your area can be relieved somewhat by taking part in an action that you feel is the right level of spikiness. It's not too scary a tactic, so it's not going to get your stress levels up. I mean, it's exciting and it'll get your heart racing but it's not a wildly militant tactic.'

'It can make you think this is a shit situation but at least people are trying to do something about it. When that's the case – there's hope. If no-one's doing anything, there isn't hope, and then that really is despairing for one's mental health and sense of possibility.'

A climate activist friend of mine in Mexico City, however, advises caution. 'I really respect what they're doing,' he tells me, 'I mean, it's amazing and in a way I wish I could do it. But the political atmosphere here and the level of awareness around climate change isn't at a point where people would make any connection between deflating tyres and ecology.' It would all be taken personally, not politically. The papers would have a field day labelling the activists as nutters and the police would likely have the political backing to be brutal. Already in London, UK, activists have been warned of jail time.[14] On Vancouver Island, Canada, police have said those caught might be given a $5,000 fine.[15] So this is a risky business, wherever you are. To date no-one is on record as having been caught, but still, it is worth weighing up the potential costs, as well as how it is likely to be received where you live. In other words, strategise.

A story was related to me recently of a conversation between a German and a Palestinian activist. The German was suggesting that the Palestinians resisting the Israeli military (IDF) might use tripods – three large poles tied together so they make a tepee shape, a raised nest that an activist could sit in to block roads. German climate activists and others have been using these very effectively for several years, largely because it is difficult for the police to get people down without the risk of hurting them. The

Palestinian activist smiled. That would be useless here, they said. The IDF would just ram them with their cars. Context is everything. Some Palestinians use a context-specific tactic that is essentially equivalent in effect. They put a backpack in the middle of the road with some loose wires poking out but nothing inside. The military stop their advance fearing that it is a bomb, coming to a standstill whilst they call in the bomb squad. All it takes is an empty bag and some wiring, rather than some poles and string.

Getting SUVs off city roads is obviously a good thing. 'In many respects,' the above interviewee put it, 'they're symbolic of the overconsumption in richer Global North countries, and by wealthier citizens within those Global North countries.' They emit far more than conventional cars. They are more likely to kill others in crashes. Those behind the wheels of SUVs are statistically proven to be more aggressive drivers. These cars also give off crazy amounts of air pollution, both from fumes and tyres. The movement's aim is to make it impossible to own a 4x4 in the world's urban areas.

But it doesn't end there. Our cities need to be systematically redesigned so that streets are safer, there is more shared open green space and widely available free public transport. How we move around our cities is key to how well – or how poorly – we function as a community. Currently the car is king. Even with electric cars, that cannot continue. Cars take up far too much space (in London, for instance, roads take up 111 square kilometres), space that could be used for walking, cycling, sitting, congregating, or planting.

Accessibility for people who need to use cars – myself included (no legs) – would have to be ensured. The spread of Low Traffic Neighbourhoods during the pandemic was overall hugely welcomed, after they had been there for a whilst, but they sometimes failed to consider accessibility needs. Overhauling how we use our space is one of the most effective ways of bolstering our mental health. We need more opportunities to feel like we belong to a community. As Massachusetts Institute of Technology (MIT) assistant professor Catherine D'Ignazio put it, 'we are designing cities that work really well for elite white men, and not very well for the rest of us'. Instead, we want what Jane Jacobs called the 'teeming city'. We want interaction, culture, places to rest. We don't want to have to pay just to be somewhere. Letting down the tyres of SUVs might seem petty to some, even disconnected from the wider goal, but it is actions like these that, together, tilt things in the direction of major change. Honestly, it's also exciting. Wear black.

C. Challenging fossil-fuel infrastructure

Any new fossil-fuel infrastructure is incompatible with a goal of 1.5 degrees Celsius of warming.[16] Following Russia's invasion of Ukraine, countries across the world doubled down on fossil fuels in the name of 'energy sovereignty', including new projects announced in the US, Canada, Germany, the Netherlands, Australia, Qatar and the UK – moves that the UN Secretary-General Antonio Guterres called 'delusional'.[17]

There are well-worn methods for climate-related occupations and sabotage. In 2008, the Camp for Climate Action in the UK saw thousands of people descend on the site of a proposed coal-fired power station in Kent. There was a massive police response and a huge amount of media attention. First, the power station's operations were delayed, then the camp meant that it could not burn coal. Later, climate camps descended on Heathrow, protesting against a planned third runway, an expansion estimated to emit as much carbon dioxide as the whole of Kenya. The debate that ensued, largely due to the camp, has snarled up progress and not a single centimetre of tarmac has been laid in the decade plus since. Similar blockades have stopped fracking projects in Scotland and slowed down plans to put in new oil and gas projects up and down the country.

In Germany, Ende Gelände (EG) has a similarly impressive legacy. Once or twice a year, EG organises mass trespasses of sites like coal mines, gas extraction platforms and power stations. They have successfully stopped projects, including the clear-cut of a forest near Cologne for the extension of a coal mine. They also occupied an open-cast lignite mine and coal-fired power station for three days in 2016, leading to the sale of the site by Vattenfall for at least 3.7 billion euros less than expected due to ecological liabilities. Others have snuck on to the site and sabotaged machinery. Sometimes this is as simple as undoing all the nuts and bolts on key bits of kit, rendering them useless.

EG is still running, though it is often a site of political tension and factionalism (one of EG's co-founders, Tadzio Müller, has often spoken about the need to improve the unsupportive and exclusionary aspects of its culture).[18] Walking under chain-link fences, through police lines and onto ground that has been earmarked for fossil-fuel development: I have heard the atmosphere around this described as 'electric' and 'liberating'.

These kinds of huge actions, however wonderful they can be, are sadly rare – though perhaps not as rare as we think. This kind of collective effervescence is very hard to find in the civilisation in which we live. Thankfully, putting a kink in the plans of fossil-fuel giants does not require thousands of people. In November 2021, two women, secured with ropes, climbed up a set of machinery and pressed an emergency stop button, instantly shutting down the world's largest coal port in Newcastle, Australia. They were part of the newly formed Blockade Australia, a decentralised outfit that now has meet-ups across the nation. In 2022, a team of kayakers paddled out into Hunter Bay and stopped coal exports dead. In the UK, Just Stop Oil pulled off a similar but smaller stunt in April 2022 when a handful of activists took over three oil terminals, one near Heathrow and two near Southampton.[19]

For a long time, indigenous activists have been at the forefront of direct actions to halt fossil-fuel development. In fact, in 2014, the Pacific Climate Warriors stopped coal ships with wooden boats in the same bay as Blockade Australia's actions in the 2020s.[20] The Dakota Access Pipeline (DAPL) has been opposed for more than half a decade. The pipeline, which was unanimously approved by legislators in 2016, led to the coming together of hundreds of tribes, mass marches and petitions, and eventually a camp at

Standing Rock on sacred Sioux land. The authorities have used firearms, sound-cannon and water-cannon. Around 300 people were injured in a single night when the authorities attacked protestors with pounding jets of cold water in freezing weather.[21] The Sacred Stone Camp that settled there was made up of 300 federally recognised tribes and up to 4,000 supporting demonstrators and activists. Clashes with the police have been common and therapists travelled to the camp to offer free services and help the community work through trauma, working alongside indigenous healers.[22] After protracted legal battles with the State, tribal members eventually won in the US Supreme Court. The pipeline, however, continues to operate. It is pending an environmental review and will hopefully end up like the Keystone XL pipeline which President Biden killed after years of similar protests.

At Migizi Camp, another pipeline demonstration in North America organised by indigenous communities, the most healing element appears to have been the birth of the community itself. Thousands have risen up against Line 3, a planned pipeline to carry around a million barrels of tar sands from Alberta, Canada to Wisconsin in the US. The camp has drawn people from all over the country, including a significant number of marginalised people with indigenous ancestry, transgender activists, a large number of support activists and even engineers who have the knowledge to safely and effectively sabotage the machinery.

Direct actions on Turtle Island (the name many have used for North America since before Columbus' genocidal arrival) over the last decade have removed the same amount of carbon as stopping 400 new coal-fired power stations.[23] Add to that the efforts of indigenous activists in the rest of the world, including the work of the Delta Avengers and Mesoamerican opposition to the Proyecto Integral Morelos in Mexico, and it becomes inescapably clear that indigenous communities are the earth's primary defenders.

Deeper democracy amongst humans can extend out wards to include Nature as an active participant. Democracy of this kind is deeply therapeutic, providing meaning, belonging, agency and purpose, as well as strong individual and collective identity. Jason Goward, a lifelong resident of Fond du Lac who used to work on Line 3, found a rare kind of communal experience when he joined the protest. An oil spill had pushed him to quit his job and step over the line dividing fossil-energy workers from earth and water defenders. 'We're building a permanent base of operations for resistance,' he told *n+1*, 'with permanent, warm structures. We're creating a cultural centre for decolonization. We are going to be fighting up here in the Upper Midwest permanently, for our Native youth's futures and for a more sustainable world.'[24]

Jason used to work in homeless shelters and support recovering addicts. He brought this experience to the movement. 'We will support anyone on the fringes of society. Native youth will find emotional support here, and physical, and spiritual, and maybe even some financial support. All four parts of medicine. You'll definitely get some cigarettes. And help with addiction.'[25]

These kinds of communities, whether or not they succeed in the headline-grabbing goal of stopping a project, are training grounds for how to foster resistance and build alternative communities. They tend to be mutually supportive as well as tactically and technically proficient; these are resilient networks for future learning and action. In 2021, Line 3's construction was completed, but the camp remains. 'We will remain an open camp, for queer BIPOC anarchists and water protectors alike to reconnect to the water, the land and the world around us,' a statement read. 'We will still be here as you sit in your homes drinking Starbucks coffee, placing bets on which football team will win next Sunday. We will still be here in the cold, in the heat, through the mud and the barbed wire. We will remain, with all of our lives, the frontline.'[26] This interconnection and commitment are crucial for planetary health, as well as for our mental health.

Andreas Malm's *How to Blow Up a Pipeline* is more history and critical philosophy than the manual that the title might imply. But Andreas also spells out some of the risks, including the stories of two US women who were found guilty of intentionally damaging the Dakota Access Pipeline in solidarity with the Standing Rock protests. After the book's initial publication, Jessica Reznicek and Ruby Katherine Montoya were given jail sentences of eight and six years respectively, and told to pay damages of 3 million dollars. Since the verdict, both women have repeatedly expressed regret for not having started more fires and done more damage to the pipeline.[27] Malm celebrates this kind of work, as does the 2022 movie of the same name. He calls for a more centralised movement culture. This is part of his wider call for an 'eco-Leninism', a top–down ecological communism. But the idea of centralising monkey-wrench actions is, by definition, incongruous with the very nature of autonomous groups. They can loosely work together on similar principles, but centralising such efforts is oxymoronic.

In *Green Desperation Fuels Red Fascism,* Klokkeblomst points out that Andreas Malm's argument rests on a disappointing analysis of a study of 27,000 ecological direct actions. Those actions, as with the indigenous actions mentioned above, have brought about significant change. We do not often hear about them, but they are happening all the time. 'Decentralised organizing, non-hierarchical networks and joyful resistance,' Klokkeblomst concludes, 'have been and will be the most effective tools to fight the builders of this ecocidal world, and to live a life free of oppression.' Deeper democracies tend to empower people far more than centralised hierarchies. Agency and freedom are important for our mental health and we need to embody them in the structure of our movements.

These tactics are centuries old. Just think of the theft, arson and sabotage of crops and equipment by captive slaves in the 18th-century United States, or the British and French Luddites of the 19th century smashing looms in defiance of early automation. The word 'sabotage' is derived from the French word '*sabot*', a kind of clumpy wooden shoe worn by weavers and used to jam automatic machinery. In the last few decades, environmentalists have been arguing about confrontational direct action, property damage and violence. Does it distract from the cause and make opposition more likely? Or is it strategic? Are there justifiable acts of rebellion and self-defence?

In Derrick Jensen's *Endgame,* a much longer and more practical precursor to Malm's *How to Blow Up a Pipeline,* the author passionately makes the case that 'love does not imply pacifism'. Towards the end of the work, Derrick talks to a military man about the practicalities of taking out fossil-energy infrastructure. They talk with a surprising amount of candour and it is astonishingly simple and safe to do using items you can easily get your hands on. My editor who claims to be a radical (I'm joking, as she knows) will not let me put the exact instructions in print, but they can be found on page 813 of Volume II of *Endgame.* Such actions are also quick. According to Derrick's interviewee: 'Your onsite time is maybe two minutes.'28

When I asked Derrick for his feelings about action and how it interacts with mental health, he was clear, but characteristically nuanced: 'I don't know that there are specific "mental health/psychological/spiritual" benefits to resistance, except that it feels good and right to do the right thing. There is a peace that comes from doing the right thing. I don't like how I feel when there is something I could be doing to prevent an injustice, and I don't do it. I'd rather be a good person than not. It's how my mom raised me.'

He then went on to say that ground actions are more about people 'achieving something tangible rather than affecting their inner state'. This is something I would like to clarify: actions like these should not be thought of as principally mechanisms for improving our individual mental health. This would just be further individualising everything, feeding into disconnection and fuelling domination. Instead, we need to see the 'tangible' achievements, as Derrick put it, and the mental-health outcomes as interacting parts of a dynamic system, a reciprocal system we step into and keep alive when we engage. These are, in turn, connected to those we meet and work with, rooted in principles of reconnection and equity, and ultimately breed their own cultural offspring. Radical as it is, blowing up a pipeline is one way of channelling that.

I was recently reflecting on Derrick's exchange with the military man when simultaneously reports of Ukrainians strategically sabotaging electricity infrastructure to cripple the invading Russian forces emerged. At the same time, right-wing reports came out claiming that the best way to spark the 'inevitable' race war in the US was to trigger a series of rolling blackouts (many believe these were seeded by Russian psy-ops). There is a double balance to strike here. First, how is the action likely to play out across existing political terrains? Could this help the movement or give dominant powers disproportionately more ammunition to attack us with? Second, do I care enough about this and am I secure enough – demographically, socially and personally – to try something as wild and risky as this? Would taking action (of any kind) improve my mental health, resilience and wellbeing, or would it likely make things worse? Could it inject new purpose and meaning into my life, or push me into paranoia and danger? Whatever the answers, is there any way of tipping the scales in my favour?

This is beautifully explored in the Icelandic-Ukrainian comedy-drama *Woman at War*. The film follows a fictional woman and her attempts to shut down a coal-fired power station, armed with only a bow and arrow and a thin line of conductive cable. Her actions are thrilling and effective, but they also jeopardise her deep desire to adopt a child. Everyone's answer to the personal and societal impacts of actions will be unique. These decisions should be made with caution. Even communicating about this stuff should be done as securely as possible for yourself and those around you. Most of the conversations I had for this chapter, for instance, were via self-destructing messages over Signal and Telegram.

D. Resistance campaigning

Because of the levels of media coverage and their relative pervasiveness, people are likely to be more familiar with campaigning organisations than with outright sabotage. Campaigners organise political lobbying, divestment movements, tech hacks, the creative disruption of events, human blockades, protests and rallies, along with a diverse range of other tactics to hold governments, corporations and individuals to account.

These organisations range from multinational NGOs at one end of the spectrum to small self-organising anonymous cells at the other. There are advantages to both. Many see institutionalisation and hierarchy as the death of radical progressivism, as famously seen in co-founder Paul Shepherd's criticisms of Greenpeace, for example. But large organisations can have great reach and they can be an effective first point of entry. Some of them also fund smaller, more radical activities and publish challenging material that ends up propelling much of the rest of the movement.

350.org funded one of the projects I was first involved in at the UNFCCC, called UNFairplay (we realised too late that, whilst perhaps a clever pun, the name was unfortunately redirecting journalists to kinky Google results). 350.org even gave us accreditation to enter the talks so that we could provide logistical support to under-represented Pacific Island Nations. Other similarly generous and decentralised organisations include the Climate Justice Alliance, Indigenous Climate Action, Indigenous Environment Network, La Via Campesina, Climate Action Network and Movimento Sem Terra. The last of these is an excellent example of how large institutions can survive for decades without their growth necessarily diluting their principles and effectiveness. Involvement in such groups is an effective and creative way to support the climate movement, as well as an opportunity to build relationships, find enjoyment in the process and settle into a sense of belonging, direction and purpose.

Other organisations provide direct support to people on the frontline. Many of us have experienced this first-hand with the proliferation of food banks since the pandemic and the cost of living crisis. This is often an ideal opportunity for getting involved with mental-health charities. I will look at peer-to-peer support in the following sections, but organisations like the Samaritans, a suicide hotline, are an amazing space to get extensive training as a volunteer and directly help people in urgent need. It is extremely difficult work. I couldn't do it, at least not yet. People who work there tell me it can be incredibly valuable, and not just for the caller:

> On the good days it all seems to come easily, [says Cerise Abel-Thompson]. But the hard days are the ones that matter most. Even when I am moping around in my own mess of life, I can be helpful to people who are at rock bottom, and to other Samaritans; it pulls me out of feeling useless. Those who I sit with in their darkest moments offer something similar. Sometimes it's the fact that they had the strength to call and ask for help, or they make a joke, or are so honest that it catches them off guard. Whatever it is, there are always moments in every shift that make me feel strong, and wonder at the strength of my co-volunteers and always of the callers.

Lots of large mental-health charities offer training in mental-health first aid and support work. It is worth looking up wherever you are, even (or especially) if you are someone who can suffer from mental-health issues. You have insights that many don't.

Smaller collectives and grassroots organisations are another interesting way to go. They generally have fewer resources, so can't afford to splash out on ad campaigns, huge installations and are – at least at first – less likely to snag coverage in the world's largest newspapers. But, and this is a big but, they have an agility and improvisational nature that makes them very exciting to be a part of. Examples include a lot of the organisations in the section above – like Tyre Extinguishers – but extend to organisations like the Bad Activist Collective, which (whilst fittingly offline for rest and recovery now) draws together artists, campaigners and activists to organise intersectional climate-justice work, whilst at the same time talking openly in written, visual and podcast form about what it means to be imperfect in these battles.[29] Some write off small organisations because they are, well, small. Others see them as breeding grounds for some of the worst macho bullshit camouflaged by right-on terminology. But, provided these groups do not fall victim to the classic problem of disorganisation and identitarian conflict (see, again, the classic essay 'The Tyranny of Structurelessness'), these groups can be immensely effective, gratifying and purposeful. They can also blossom beyond the confines of the small-scale, unpredictably and improvisationally giving birth to new realities. As Della Duncan, host of the Upstream podcast, often reminds me by quoting the Liverpuddlian Craftivist Sarah Corbett: 'If we want our world to be beautiful, kind and just, our activism has to be beautiful, kind and just.'

As a principle, I would recommend joining a group based on its character and structure, how it runs and how it feels, rather than on its particular campaign focus. Sometimes you are really lucky and you get both. For some, that might look like one of the Red Gyms popping up around the UK, spaces where lefties are trained in boxing as self-defence, anarchist, autonomous and/or socialist principles. Some of them have strong connections to unions and have even sent delegations to protect victims of deportation, stand between queer marchers and violent homophobes, as well as helping potential victims of abuse defend themselves. Solstar Gym, a women-led gym based in Tottenham, London, is one of them. Their slogan is 'training together for the struggle'. Intersectional to its core, it was founded by Dr Ella Gilbert, who happens to be a climate scientist as well as a boxing coach.

Sometimes rocking up to these kinds of spaces can be intimidating (any meeting, not just a boxing gym). I can't count the number of times I have turned up to introductory events where I didn't know anyone, only to dawdle and delay outside before heading straight home due to my anxiety. Other times I have gone in and been so detached, out of a combination of scepticism, fear and cynicism, that I didn't actually get much out of it other than feeling like an outsider. I have found it much easier to go with someone. For me, that somehow breaks the seal. If you find a group of people you would like to be part of, then do try to go to lots of their meet-ups. Consistency is key. Unless you put the time in, it won't ever feel like home.

In the UK there is a wide array of groups to try. There's Extinction Rebellion (more on them later), Just Stop Oil, Insulate Britain, Stop Cambo, Fridays for Future and more. Mikaela Loach is a health and climate-justice organiser with outfits like Stop Cambo, co-host of the *YIKES* podcast and author of the 2023 book *It's Not That Radical*. Here's what she told me about the current discourse around mental health and climate change:

A big part of what comes from climate dread and climate anxiety comes from a feeling that you can't do anything about it. There's an unmovable thing. It's only going to get worse. That's why I think one of the best antidotes to climate despair, or whatever you want to call it, is climate action. We'll continue to feel this way unless we do something about it. Often I see the responses to climate anxiety being like: 'Oh, just go for a walk; just switch off; or don't read the news.' That's only going to make you more anxious, it'll only make things worse. It's not making anything better. . . .

A lot of these problems are the consequence of a society that is exploitative to each other, a society that makes us unhealthy to each other. It's because of capitalism, it's because of white supremacy, it's because of these bigger issues. So, the best way that we can tackle climate anxiety or climate dread is to get involved in movement organising. From my perspective, I know that I was a lot more climate anxious before I got involved in organising. I realised that we can do something about this. We actually can. We have so much power and we can create a lot of change. We already have solutions that are not just about stopping the world from getting terrible, but that will make things better overall. You only really realise that, and recognise the power and the agency that you have, by actually doing it.

Perhaps the most famous recent action by one of these groups was Just Stop Oil's Phoebe Plummer chucking a can of tomato soup on Van Gogh's *Sunflowers* at the National Gallery in London. The painting was covered with glass, but the action still caused international uproar. These groups aren't all radical direct-action zealots (although I do love a radical direct-action zealot in the spirit of Roosevelt's lunatics). They have also been piling on pressure using social media with massive success, organising petitions and letter-writing campaigns, and made the news by interrupting big public events like the prime ministerial candidates' debates, high-level ministerial meetings, steadfastly refusing to go to school and blocking entry to the AGMs of major oil companies. They are responsible for hundreds of public events across the country, too. These are the building blocks of both climate awareness and climate action.

I am still waiting with bated breath to see an upsurge in climate hacktivism. There was a dramatic spike in young people learning to code during lockdown and we have seen a few imaginative applications of this knowledge in the activist space. My favourite recent examples both come from a software developer with the TikTok alias Sean Black (@seandablack).[30] He wrote one script that semi-automated fake job applications to Starbucks, a protest against their anti-union stance that meant the company was flooded with more applications than it could sort through. Another similar stunt saw Black disable a 'whistleblower' website set up by Texas Right to Life. The site was intended to snitch on women who were suspected of flouting Texas' new anti-abortion legislation. Black's software allowed anyone with an iPhone to generate a false report from a Texas zipcode. The tips were obviously fake, including Marvel characters and a man who wanted to abort his thirty-year-old son who wouldn't move out, but they caused the site to crash repeatedly. We have got enough people trained up now to mess with Big Oil in a similar way, snarl up the production processes of dirty goods and hamper politicians who fail to take the crisis seriously. We just need to get creative, like the anonymous hackers who wrote guidelines in the now delightfully vintage and still beautiful 2004 edition of *Hack This Zine* (available free online).[31]

E. Logistics and support actions

Without logistical work and support, action movements die. Often this is because of burnout. Often it is because of an ever-growing, almost pyramidal organisational structure, where people who are in trouble need help and there are not enough other people with the energy to support them effectively. At the very beginning of 2023 Extinction Rebellion (XR) released a statement saying 'We Quit'. Later, I will investigate how to foster more regenerative and durable movement culture, so that groups do not flash up and burn out every three years. But for now, the focus is on the importance of support. Part of the reason XR buckled and had to change tactics was reportedly because so many of their resources were spent supporting people who had been arrested. Lawyers, emotional support, practical support for those in prison – this takes a great deal of work. Energy and resources run dry as the number of arrestees and detainees increases and the length of time they are incarcerated stretches out. There has been a lot of analysis of this and it seems to be that XR's dwindling represents an age-old pattern of a core of super-dedicated members, loosely supported by a larger number of satellite individuals who became less and less enthusiastic and felt they had little agency.[32]

Just Stop Oil's actions have seen an ostensibly different process emerge (although we may just be in an earlier phase of the same thing). Phoebe-sometimes-I-throw-soup-Plummer[33] spent a month on remand and was subsequently released, but at the time of writing there are around twenty climate activists still behind bars in the UK. Recently, some of them told Euro News just how important it has been to receive postcards and letters of support, as well as positive emails and social-media activity. Marcus, who climbed up the Queen Elizabeth II bridge to shut down a motorway, called this contact from outsiders a 'lifeline'.[34]

People have tried to provide other kinds of support. Tipping Point have been offering help, fostering nourishing organisational structures and grassroots structures, and supporting people within existing groups, both of which have strong mental health aspects. What makes people want to join, feel heard, and stay? Tipping Point were involved in work at a camp at XR Greenwich, the atmosphere and culture of which one activist described to me anonymously as 'toxic'. He continued: 'How do we create groups that are nourishing and how do we support people in those groups that might not always have nourishing experiences, because we live in a world with various different traumas and people will trigger one another and hurt one another unintentionally. We need to improve the culture of movements themselves, they can be damaging and draining, but external organisations helping groups like XR can be a catalyst for change, as well as providing direct support in difficult situations.'

The therapists who went to Standing Rock, mentioned earlier in this chapter, fall into a similar category, as do organisations like Mani and TEAP, which featured in the Nigeria chapter. So too do the mental-health support trainees in Dr Asma Humayun's network in Pakistan, along with the psychological and logistical framing offered by Fahad and Emiliano's organisations in Pakistan and Mexico.

This kind of work also includes cooking food at camps or protests, filming, keeping people company who have glued themselves to stuff, training and then practising as a legal observer, platforming each other, offering technical advice, making introductions, fundraising, building websites, leafleting and more. The movement is made up of many parts. They interact like a complex ecosystem. Stepping out of the capitalist bent of the hero narrative and into a more comfortable and nurturing alternative vision of collective flourishing will go a long way towards us all feeling like we matter.

My grandfather was a historian of science, a Darwin scholar. He never got bored of telling me how completely capitalist society had abused the theory of evolution. Herbert Spencer was a social Darwinist, a man who twisted Darwin's biological research into social and political theory. He was actually the person who coined the term 'survival of the fittest'. Today, virtually everyone takes 'fittest' to mean 'strongest'. Darwin adopted the term in the fifth edition of *On the Origin of Species*, but he was adamant that 'fittest' actually meant 'best fit'. We all need to fit into the ecosystem that is the human struggle to avert climate chaos. Some will find their niche in logistics, others in arrestable actions, others by writing. Herbert Spencer can do one.

F. Art-tivism

Whether you are artistic or not, getting involved in art-tivism is a lot of fun.[35] It can mean running around in the dark and climbing up scaffolding to put posters on a roof, meeting people in a bustling studio, honing a creative craft and putting it to good use and, fundamentally, looking at the world in a different way. Art done right and put in the right spot can open portals into other worlds. Visuals can be powerful influences. We all know advertising is powerful. We can subvert its tools so that our messages can be absorbed just as instinctively – only to spur action instead of consumption. There is a whole sub-genre of art-tivism known as subvertising or brandalism (the community is keen on portmanteaus). Subvertising, which you will probably have seen somewhere if you live in a city, is any attempt to use the iconography of the oppressor against them. A rash of fun and beautiful subverts hitting out at BMW and Toyota spread across Europe in Jan 2023. One of the people involved told me:

> The art of editing, tinkering with, humorously changing advertising is as old as advertising itself. What groups like Brandalism, or Special Patrol Group or the Subvertising International Network do that is particularly fun is make your own artwork and put that back in the space. That can be creative and colourful and very directly empowering because you access those corporate advertising spaces that are reserved . . . for those with the money to pay to put their message up, and you regain that space.

You're subverting advertising in an act of jujitsu, using your enemy's weight against them. You're using the brand recognition of their logos that they've carefully crafted using millions of dollars and dozens of advertising agencies over many decades, [sending] all of that back at them by attaching a different meaning to that brand. It's fun. It's a bit edgy. It can make you smile, bringing a bit of light relief to a dire situation.

Subverts can take the form of mock adverts, like airline Ryanair having their name changed to Ruinair (or Rienair in France), brightly coloured SUV adverts with slogans like 'Ignore the Kids, Burn the Planet', or posters like the one of the Deepwater Horizon disaster pasted with the words 'Shell Chats Absolute Gas'. You do not have to design these yourself. Many of them are in the public domain. There are also online guides, like *The Subvertising Manual,* which has a detailed guide on design, how not to get caught and even advice on where to get keys to open closed ad spaces (the keys are used pretty widely globally).[36] As the anonymous interviewee above also told me: 'People have a great time and express a sort of eureka moment when they get hold of the key and somebody shows them or they figure it out themselves or maybe they watch a Youtube video or pdf explainer . . . you just slot the allen key in, give it a quick quarter turn and the bus-stop advertising cabinet pops open and people are like: "Oh, is it that easy? Brilliant."'

Art-tivism ranges from low-key underground acts of graffiti and postering, all the way up the scale to stunts like the inflatable Trump Baby, the huge projections onto buildings by Led By Donkeys in the UK, or the moving '99% Bat Signal' during Occupy Wall Street. Art-tivism can include photography, Photoshop, graphic design, music, video mash-ups, memes (yes) and performance art; from stage theatre to TikTok comedy. Whatever it is, get good and get seen.

As with many fusions, this kind of art and activism can sometimes be the worst of both worlds – neither beautiful/moving nor effective messaging/motivation. It is helpful to explore some of the most effective protest art for inspiration. I personally love the work of the UK-based Autonomous Design Group who have posters you can buy and paste all around town.[36] Whether you are putting up someone else's designs – an act of rebellious artistry in itself – or making your own pieces, you can choose to be as open/safe or underground/risky as you like. You can print flyers, you can make campaign material for social media, you can stick posters on a polluter-funding bank's windows. I use wallpaper glue, which is almost as cheap as home-made altern-atives. Take a roller or a squeegee, and scout the area before, to make sure you know your way around. It is best not to be doing this alone, either.

Graffiti requires fewer materials, but can take a bit of practice. A nice stencil made of thin, solid card will go a long way to saving you time and having a consistently striking result. You don't have to be Banksy or Foka Wolf to make it worth it. Stencils can also be used for what is called 'negative graffiti'. This is the activist's version of writing 'wash me' with your finger on a dirty van. It works best on a grimy surface (a pavement, a wall, a bollard, etc), where you whack down the stencil and apply a pressure washer. Pressure washers are not easy to get hold of, but some activists claim a toothbrush works. I think you would need to spend quite a bit of time with a toothbrush on the pavement to make an impression but you can give it a go.

I love doing shit like this. I always have and I hope I always will. Some might think it is mindless vandalism but it's really not. It is a way of having a different relationship with where you live. If you make an effort for it to be beautiful and you want it to help the people who see it and the planet they live on, then how does what you make have any less right to be in public view than a billboard advert? Find a community, explore your world, beautify the resistance.

9

RECONNECT: how to get together

'Action on behalf of life transforms. Because the relationship between self and the world is reciprocal, it is not a question of first getting enlightened or saved and then acting. As we work to heal the earth, the earth heals us.'

Robin Wall Kimmerer

'You cannot transform the society of people if the people are not part of the change.'

Angelique Kidjo

RECONNECTION IS ABOUT MORE than connecting with your inner self. It can be about that, but it is also about spinning your sense of connection out into your wider social and environmental situation through relationships and action. Reconnection is about consciously embedding yourself in the world again. It is about strategically increasing your engagement, along with those of others and increasing your power and impact. If this is done relationally, which much radical activism aims to, then what we are talking about is increasing meaningful connections between strangers and deepening those with others. It is also about shared empathy, critical thinking and perpetual exploration. All of this, fortuitously, happens to be both excellent movement-building strategy and a reliable route to psychological resilience. Our historical experiences of trauma and post-traumatic growth can increase empathy and the ability to recognise emotions, both of which help in supporting others.[1] Building relationships on a more equal, mutually supportive footing is one of the things that makes rebellion durable – and enjoyable.

Working relationally is a core tenet of community organising. We are all familiar with the idea of canvassing, but few have heard of 'deep canvassing': a methodology that uses empathic dialogue to connect issues and campaigns with people's existing hopes, fears and lived experience.[2] This is also about learning from the people you are speaking to, having an actual conversation rather than just ranting at someone. No wonder people are so disengaged politically if the latter is the norm. Deep canvassing is effective at building relationships and commitment, but it can also change people's minds in a way that conventional canvassing cannot.[3] It also changes the presumed structural dynamics of activism. Deep canvassing and its associated approaches meet people where they are. They give agency to everyone involved. New people might even want to join in. This plays to our advantage, building the power of movements strategically through radical inclusivity.

Receptiveness and democratic discussion are hard to teach. Humans are evolutionarily wired for co-operation, to the extent that some psychologists have dubbed us the 'ultra-social animal'.[4] But we have been born and bred in hierarchy, encultured outside of communal models and into one of stratified power imbalances. No matter what Jordan Peterson says about lobsters, we are not primarily programmed to dominate and be dominated.[5] That kind of relating to each other is soulless.

People have a very malleable conception of what 'community organising' is. Differentiating between regenerative rebellion and a more paint-by-numbers approach is difficult to do unless you are familiar with some emancipatory precepts. The US government has even used this blurred definition to its colonial advantage, offering a free online course in community activism on its page dedicated to the Association of Southeast Asian Nations (ASEAN). Don't take it![6] (They also have one for their

Young African Leaders Initiative, with a sneakily tiny logo.)[7] The online Organising School from the organisations Changemakers and Tipping Point is much, much better. It also includes information on safeguarding our mental health as activists seeking transformative change.[8]

Saul Alinsky's imperfect (but useful) list of twelve *Rules for Radicals* is a good start as a guide. A famous radical campaigner and theorist in mid-19th-century America, his rules include: 'a good tactic is one your people enjoy', 'power is not only what you have but what your enemy thinks you have' and 'wherever possible go outside the expertise of your enemy'.[9] Alinsky was a tactician and a community builder. He founded the Industrial Areas Foundation (IAF) in 1940, one of the many surviving training centres around the world to coach people in these more complex, interesting and effective models of activism. The IAF focuses especially on training working-class communities. Marshall Ganz, who left Harvard in the 1960s to join the Civil Rights movement and later orchestrated much of Barack Obama's presidential campaign, has published many detailed yet accessible models of organising. Perhaps the most prominent, *Organizing: People, Power and Change*, includes tenets like: 'build intentional relationships as the foundation of purposeful collective action'. The entire text is available for free online.[10] From anarchist organising with Emma Goldman and the Industrial Workers of the World in the early 1900s, through to Martin Luther King (MLK) Jr's Southern Christian Leadership Conference and Weather Underground in the mid-1960s, to modern abolitionist movements, the Sunrise Movement and Bernie Sanders' barnstorming strategies today, rebels in the US have long been innovators in developing diverse techniques for ushering in change. Then again, maybe they are just the most widely circulated, consistent with Global North hegemony.

Anti-colonial struggles, worker rebellions and outright revolutions have a long history and a diverse, thriving legacy. One outfit, Slum Dwellers International (SDI), is a decentralised community-organising institution with millions of members spread across thirty-two countries in Africa, Asia and South America. They build 'federations of the poor,' empowering local communities through their own labour and collectively advocating for systemic change.[11] SDI has a strong stance on climate change, insisting that: 'Organised communities have the skills, capacities and systems to drive and deliver locally led adaptation to the climate emergency and channel climate finance to those who need it most.'[12] The people who live in these communities know best. In India, SPARC and Mahila Milan (Women Together) use decentralised community-organising techniques to support pavement dwellers, offer credit and meet people's housing needs.[13] Indigenous movements in South America often use models similar to but usually predating community-organising principles, including successful opponents of extractivism and defenders of the Amazon rainforest in Peru, Brazil and Ecuador.[14] South Africa's Centre for Environmental Rights has training in community organising, having worked effectively with health campaigners and people defending land against mining, as well as providing activists with legal support.[15] The thread that runs through all of these efforts is relationality – reconnection on the basis of equity. It is the inverse of the disconnection and domination into which we have been encultured. It is also where you find new ideas, build power and generate momentum.

Speaking of momentum, community organising played a big part in Jeremy Corbyn's surprising ascent in the UK. The Labour Party's Community Organising Unit (COU) was a key driver behind many of his unexpected successes.[16] He lost two elections, but one of his greatest legacies, in addition to making progressive ideas popular with and palatable to millions, is the rich ecosystem of community organising that survives across the country today.[17] This infrastructure and culture is fundamental to maintaining and strengthening progressivism. Corbyn's successor, Keir Starmer, disbanded the COU, expelling many of its activists. The party's membership fell by almost 100,000. In general, many people leave activism because of burnout and despair, but a huge number leave because of a simple lack of agency. If you are not being listened to, what's the point?

We need to think about movements in a fresher, more enticing way. They should be about building community, engaging in decision-making as respected equals and a constantly evolving process of creating the future with people you care about. A march, for instance, should be about more than just turning up and going home. It is the middle portion of a long process that might involve your helping to organise the event, meeting people there and, if you are smart, bringing them into the movement and offering them training. There is immense expertise and knowledge in every person. We have to start from that and, with humility, love, hope, faith and critical thinking, spin out from there.[18] As Gandhi definitely didn't say, we need to: 'think local, act global'.

Power mapping
Power analysis is a core element of strategic campaigning. Part of this is learning about the systems we live in and our place in them. Sometimes drawing it helps. A power map is a useful compass. What we want to do is place individuals, organisations, institutions and media somewhere on the map. Where they are on the horizontal x axis describes how much they agree with our objective or position. The vertical y axis tells us how influential they are and how much power they have over the thing we are trying to change. To do this well, it helps to expand our knowledge by doing research and talking to people familiar with the space. That might sound daunting, but it is something that can be done with others, an exercise that turns into a form of movement building if we approach people with a collaborative mindset.

Later, we can draw up a diagram of each actor's relationship to each other. The organisation Little Sis (as opposed to Big Brother) hosts an open online database of powerful people's relationships to each other which might help, although it is quite US-focused.[19] The diagram below comes from the Beautiful Trouble online toolkit, a free treasure trove of activist guides, tactics, methodologies and theory.[20]

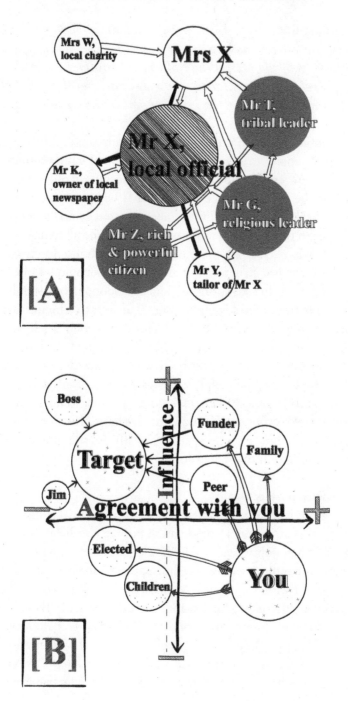

[A]: Relationships map: black arrows indicate accountability to, while white arrows indicate power over. [B]: Elaborated power map: with 'you,' the 'target' and others, all positioned according to their level of agreement with you/your cause, and their level of influence and power over the situation you are trying to change.

These can get really complex – but also really pretty. Little Sis, for example, made an amazing visual depiction of who is funding the Dakota Access Pipeline and their relationships to one another which can be found for free online.[21] It is important to understand and acknowledge the importance of strategy and a couple of the techniques in our toolbox. We don't need to use the master's tools exclusively, but too often decentralised activism can eschew not just structure but also strategic techniques of any formalised kind. It feels more organic and spontaneous (reminiscent of Jez from *Peep Show* shouting: 'I answer to a higher law, the law of "if it feels good, do it"'.') This can become a self-generating system, but more often either diffuses into nothing or rachets up into burnout territory. If we have an idea of where we fit in the system, the people we can collaborate with and how best to use our energy then that gives us more time to rest, recuperate and reconnect. It's more fun. It's more effective. David had a plan when he went up against Goliath. A little pebble and a little decapitation made him king.

Local community action
There's a tendency on the Left to wax lyrical about this kind of organising. That is okay in moderation but we also have to walk the walk. When Bobby Seale of the Black Panther Party was asked if he saw any conflict between socialism and community organising, he responded with a flat 'yes'. 'Most of the Socialists would just talk. They intellectualized the crap too much,' he said. 'Basically, our ideology is community control.' 'But that is patchwork Socialism!,' argued Seale's 'intellectual' critics. 'Patchwork my ass,' he replied, 'That's necessary. [We need to be] evolving a system and an economic practice of fair-shares equality and fair-shares

access. It's from the grassroots up. Not some pig pile structure down.' Seale travelled up and down the West Coast in the seven months between the assassination of MLK and Nixon's election in 1968. He sought out left-wing students at universities, talked to them and gave them a mission. As he reported it: 'I gave them a constant theme: You are organizing around programs. You want programs in the community. You will put up a breakfast program. And once you get through that, you want to put up a free medical health clinic. [. . .] I said, any other programs you can come up with after this, do that. Do that process.' By the end of his trip his organisation had grown from 400 members to 5,000.[22]

Organising has to be pragmatic. It has to address real needs. A collaborative approach to solving immediate problems provides an oasis of communality, nestled within the neoliberal desert. We have seen an immense amount of this spring up in recent years, in the form of resisting deportations, renters' unions, community safety, strike action, sharing food and shelter and more. When people support each other, become empowered and have the opportunity to expand their critical thinking and action, a qualitative shift occurs. It is a refuge, but also a promise of much more: a different way of living.

I met Ola Adéyémi, a nurse, in a psychiatric hospital when I was a patient there. I had insomnia and she kindly took me out in the middle of the night for a cigarette. We talked about systemic drivers of mental-health issues, covering climate change, inequality, racialisation, colonialism and capitalism. 'Mental health,' she put it bluntly, 'is social.' Ola grew up in northern Europe and on moving to the UK quickly noticed the huge difference that economic precarity made to mental health. She decided to run a theatre workshop on mental health in a largely Black and minority ethnic (BME) community and saw that everyone had some connection to mental-health issues. The community needed support, but was not getting it. Ola knew that BME people not only had higher rates of serious mental illness, but were also less likely to receive government support. Even if they did, they were significantly more at risk of being manipulated or abused in care.[23] Instead of just educating and connecting people with existing services, she and her fellow organisers set up a stationary bus where people could come and chat to volunteers about their mental health. It was liberating. Some people who initially came for help ended up contributing to the project. They also set up a programme in local barbers and hairdressers where customers had the opportunity to open up in a safe space.

Similar projects exist in London today. Black Minds Matter connects Black people with free mental-health services.[24] The Black Men's Consortium runs workshops on mental health in Brixton, London. They have also released a few podcasts including their members talking about how they look after their mental health, seeking to educate but also to break with unhelpful expectations of strong and silent stereotypical machismo.[25] Covid dealt a blow to many of these kinds of projects: the Refugee Council, for example, have had to shut down a lot (but not all) of their programmes offering free mental-health services to displaced people.[26] Listening cafés like those offered by Jennifer Uchendu's organisation in Nigeria (see Chapter 4) are a good model if you want to start something of your own.

Other peer-to-peer support models abound, from informal online chat rooms to dedicated sites like Koko, a non-profit portal where users can get instantly involved with peer support (seeking or offering help at the touch of a button), take self-help courses, develop a safety plan and use crisis lines – all for free. Koko has interacted with over 2 million users and almost three-quarters say they feel better after spending some time on the site, both by helping and being helped.[27] I have spent some time using it and for an anonymous online platform it's actually pretty good. After my most recent session, someone sent me a thank-you note after I had chatted with them about their depression and a family sickness. It immediately reminded me of something a friend told me about the power of helping others. He is a member of Narcotics Anonymous (NA) with years of clean time. NA is a deeply rooted programme benefiting from decades of shared lived experience and expertise. A lot of NA recovery is based on service, on talking to and actively supporting each other. This friend shared: 'I've felt more elated from helping other addicts than I ever did from Ecstasy. It's not rushing, it's not a dramatic come-up, but it's this feeling of acute okay-ness . . . I think the best word for it would be love, corny as that might sound.'

Face-to-face support groups can be relatively easy to find. It is also possible to set this kind of thing up yourself, either part-nering with an organisation or going solo. The Community Tool Box has a pretty comprehensive guide you may want to use.[28] This might sound like the mad leading the mad, but if you have the right safeguarding measures in place then there is nothing wrong with that. In fact, it can be beautiful. If you are looking for people who 'get it', then the collective support of like-minded people is powerful medicine.

Climate action can have a similar bonding effect. Community organising to directly help people in your area builds resilience and belonging. It might be a good start to find a mutual-aid group in your area to have as an anchor. Getting creative helps. If you live in an area of huge inequality, as I do, you could try going door-knocking in the richer neighbourhoods, asking for cash to support those struggling in the community. If you can engage in a bit of deep canvassing whilst doing so, all the better.[29] In these dark times you will probably be surprised at how much money some rich folk are willing to give to support an earnest progressive project in their own backyard. You can also raise cash through online crowdfunding for an event or for small donations.[30] You may even be able to get council funding. Obviously it's not all about the money. Then again, though I hate to quote Ye (formerly Kanye West), 'having money isn't everything, not having it is.' With cash you have the power to equip a community-organising campaign with the materials to insulate people's homes, install community-owned renewables, institute an accessible Low Traffic Neighbourhood, or put money into a hardship fund for the unhoused, people struggling with their energy bills or those going without food.

In modern Britain we have nearly a million children living in food poverty.[31] One in four parents skip meals so they have more to feed their kids.[32] One incredible response to this kind of unnecessary and brutal precarity is the movement for a National Food Service (NFS). They say no-one should go without food, a hard argument to disagree with, and in some of the cities where they operate, NFS also provides space for people to gather, bring ingredients if they can, cook together and eat together.[33] This is not just handing people food parcels, it is connecting the community.

Other potential allies include more conventional foodbanks, a vital lifeline for millions, as well as renters' unions – like the London Renters Union – progressive labour unions, organisations like Transition Towns and wider efforts to propagate more connection in communities. This combats loneliness, destitution and despair. But building community has surprising emergent benefits of its own. As *Guardian* journalist and author George Monbiot puts it: 'Turning such initiatives into a wider social revival means creating what practitioners call "thick networks": projects that proliferate, spawning further ventures and ideas that weren't envisaged when they started. They then begin to develop a dense, participatory culture that becomes attractive and relevant to everyone rather than mostly socially active people with time on their hands.'[34]

This movement needs to be a mass movement: an interconnected, ballooning network of self-sufficient programmes. You can't just build it and expect people to come. It has to be attractive (and not in a 'maths is cool, kids' way). It's not a chore, it's liberation.

Launching legal challenges

Existing legal infrastructure is often an effective tool of the oppressor, as we know from the recent tightening of restrictions around climate activism. As Martin Luther King Jr put it, 'One has not only a legal, but a moral responsibility to obey just laws. Conversely, one has a moral responsibility to disobey unjust laws.' That said, there are quite a few ways to use the law to our advantage, often by finding chinks in the armour of jurisprudence then prising them open with a long lever. The Good Law Project is doing just that. Based in the UK, they are probably best known for repeatedly suing the government over corruption and a serious lack of transparency. But they also launch cases about transgender rights, racial discrimination and climate change. In 2020, they forced the government to change national energy policy, relinquishing new oil and coal power generation and scaling down gas projects.[35] A year later they sued the government over inconsistencies in its Net Zero Strategy, and won.[36] As Good Law Project founder Jolyon Maugham told me: 'We believe the real power of climate litigation is that it, in its bringing and supporting, provides a kind of personal agency in an area where there is so little. So, yes, we are funded by everyday people . . . But also, we need communities to gather evidence for us and suggest litigation and share our messages.' Similar organisations include ClientEarth, who have 168 cases in process and operate in fifty countries. One of their projects is in Gabon, where they work with local lawyers and communities to protect land rights and access to primary forest, as well as ensuring that any new laws are gender sensitive, so women can be involved in forest governance by design.[37]

Activities like these are hard to engage in from a standing start, but lawsuits are usually much stronger if there is a social movement backing the cause. The Proyecto Integral Morelos in Mexico (see Chapter 6) is a vivid example of community organising and litigation going hand in hand. Imagine a legal challenge has been filed against a coal mine. Solidarity action could involve marches, direct action, sit-ins, deep canvassing and targeting politicians. It would be a good idea to contact the law firm involved before getting going, as it might hamper their efforts for some reason of which you are unaware.

Collaborating to strengthen mental-health lawsuits can be more straightforward, even if more emotionally challenging. If they are based on suing a psychiatric unit or a particular doctor they are often site-specific. Site-specific community-based actions can hit hard. Mental-health suits usually involve families – people already intimately connected to the plaintiff – and can result not just in retribution but in wide-ranging changes in ward policy or even national regulation. The legal side of these fights is important, but community organising to redress injustices – often negligence, personal abuses of power and outright abuse – have a significant impact on the behaviour of medical professionals who are used to operating in relative obscurity. I wrote part of this book in a house built and lived in by the renowned Mexican artist Feliciano Bejar, who died in 2007. Feliciano was a famous artist who campaigned on ecological issues, mental-health reform and gay rights, as well as painting dreamlike scenes. He also made beautiful glass sculptures that refract your vision so that it is like you are looking through a portal. Feliciano was once a patient in a psychiatric unit owned by military medics, following a breakdown. His partner Martin Foley tells me that it was environmental activism in particular that drove him to despair. Feliciano said he was abused in the hospital and on his release he used his platform to publicly sue the military doctor who had forced treatments on him. This sprawled and became a campaign to destigmatise mental illness in Mexico and shut down the country's outdated and oppressive psychiatric hospitals. Some of the most brutal places were indeed shut down, Martin tells me, following a concerted union of social movements and legal battles. The battle was also, it must be said, extremely painful for Feliciano and his loved ones.[38]

At any scale, community organising around mental health can mean meeting people with similar experiences and stories. Campaigning together relationally creates strong bonds. This could potentially generate an informal (or formal) support group. The most valuable thing I gained from my time in psychiatric units, other than temporary safety, was meeting the other nutters in the nuthouse. Many are still my dear friends. We are often each other's first port of call in times of crisis. There is a deep solace from being so intimately connected to each other, people who understand so many of each other's struggles on a personal level. Versions of these spaces already exist online and in person, like #PsychosisChat on Twitter and the groups facilitated by the Hearing Voices Network.

A bold way to engage with legal challenges is to launch one. This is hard, takes years and can cost a prohibitive amount of money if you do not have a financial backer. Some have done it anyway. *Juliana v. United States* is an ongoing case where a diverse group of young people are taking the US government to court for knowingly failing to ensure a safe climate for them and future generations. They are asking for 'a declaration of the federal government's fiduciary role in preserving the atmosphere and an injunction of its actions which contravene that role'. Thousands of people have rallied in support. The case is ongoing.[39] There have been more than 2,000 climate cases worldwide, the vast majority in the US (they're a litigious bunch). Other high-profile suits include a group of Colombians aged between seven and twenty-six who sued their government over climate change and the deforestation of the Amazon. They won. The ruling forced the Colombian government to deliver a net-zero deforestation strategy.[40] People have sued on the basis of human-rights infringements, the failure of governments to adapt to climate change and, in Chile, successfully demanding that the phasing-out of fossil fuels incorporates a just transition (specifically, finding jobs for workers after a fossil-fuel plant was shut down).[41] It might sound silly to sue the government over climate change. We might as well sue God. But little did you know (or maybe you did) that in 1970, Russel. T. Tansie did exactly that on behalf of Betty Penrose, after her house was hit by lightning and burned to the ground. Tansie accused God of 'careless and negligent' control of the weather (sounds familiar).[42] When God failed to turn up in the California court, Penrose was awarded $100,000.[43]

Change in your workplace, school or places of worship
Organising in spaces that you know can be nerve-wracking. It is quite easy to be labelled as the 'eco one', inviting attacks from climate sceptics and people who think it is funny to lay into the wokerati. But it's an amazing way of getting to know people on a different level. I set up a green council when I was in secondary school. We pushed for the standard stuff: recycling, solar panels on the roof, education on climate change as part of the curriculum. It was tough, and I hadn't yet learned about

community-organising principles, but we were good at it (we even won a national award). It was my introduction to activism proper and it lit a fire under me.

Changing the lightbulbs[44] in the office might seem inconsequential in the grand scheme of things. What is important, though, is the relational process you go through to get things like that done. It is the relationships that matter: connecting with people on a deeper level, finding out who they are, what they need and how they care about important issues and the process of change. If organising of this kind can be done in a regenerative way – not just standing up and giving a speech, and definitely not shaming people into recycling – then a healing and supportive dynamic should take root. People will start talking to each other in a different way, seeing one another as more than just colleagues and making demands of their bosses for a different energy provider or mental-health support. It is far easier to do this as a group, which is one reason unions are so vital. The hard part is stepping over the boundary from invisibility into a potentially controversial role but it can start over a beer at the pub, or a note slid across a desk. Just because you are trying to catalyse change it doesn't mean you have to jump up and down and make yourself a target.

If you practise an organised religion, you will likely be very familiar with the charitable side of intentional community. Religious people are, in general, more likely to donate to charity and to actively volunteer.[45] Interestingly, though, religious belief is connected with higher levels of volunteering in every area *except* environmentalism.[46] In Abrahamic religions this might have something to do with the notion of humans being made in the divine image and having a responsibility to 'subdue' the earth. But there is a long and vibrant tradition of spiritual environmentalism. Liberation Theology, for instance, interprets Christianity as a path towards the salvation of the oppressed through political and ecological freedom.[47] There's also Pope Francis' 2015 ecological encyclical, which many took as both a radical ecological tract and a systemic critique of capitalism.[48] Jesus was a rebel. So was the Buddha. So, some say, was the Prophet Muhammad. According to the scholar Fazlun Khalid, the Qur'an is so implicitly ecological, that the term 'Ecological Islam' is a tautology. This is made clear in scripture like: 'The creation of the heavens and the earth is far greater than the creation of humankind.'[49] Judaism, Buddhism, Hinduism: each have their ecological branches and acolytes. Marrying these with the charitable infrastructure of a place of worship is demonstrably doable. What might be harder is moving from some more transactional charitable models towards community organising and mutual aid, though many faith-based examples of the latter exist. Godspeed.

Filling the gap: getting elected

The Alt-Right's rise to prominence in the US and the continued influence of groups like the Tea Party are in no small part because they are being super-strategic. I am not talking here about the Koch brothers and Peter Thiel, although they have a lot to answer for. Nor am I talking about Q, Alex Jones or the Proud Boys. I am talking about individual citizens' successes in getting elected to local government.

Since the 2020 US election there has been a proliferation of recall campaigns for school board members, right-wing individuals standing in their place and winning elections to educational bodies. It is largely propelled by a mass mobilisation of previously apolitical parents. Two of the organisations pushing this are No Left Turn in Education and the shadily funded outfit Parents Defending Education. Both manipulate messaging about critical race theory, 'radical indoctrination' and grooming, pushing parents into citizen activism using fear. In its first year, No Left Turn established seventy-eight chapters in twenty-five states.[50] Parents Defending Education has an interactive 'IndoctriNation Map' listing crises to address across the US. It is scarily detailed, somehow including pdfs of local school lesson plans and present-ation slides, information about 'incidents' (like curricula supporting 'LGBT pride' and 'culturally responsive practices') as well as invitations to existing parent organisations to join. Somehow, this has charitable status. The Tea Party Patriots website also has a page dedicated to training parents in attending and disrupting school board meetings and taking over school boards. Their mission is clear: 'If we don't take control of school boards across the country, we will be left with a future that hates our country and each other. It's up to us to make sure that doesn't happen!'[51]

They are succeeding. It might sound boring, sitting on a school board or organising a protest, but ... well, the children are our future, etc. The progressive left needs a response. We have higher levels of participation in community activism, but a lot of people on the left, myself included, sometimes focus too much on macro-level policy or local campaigns that are detached from local politics. This leaves a wide-open space for the Right. In the US, they strategically turn up at town-hall meetings, get their voices disproportionately heard at local political consultations and run for council seats. The Republicans have famously held several recent record majorities in state legislatures across the US. The surprise success of the UK Independence Party (UKIP) in securing Brexit was largely due to its dual identity as a movement and a fully fledged political party.[52] UKIP only overtook the British National Party after taking a leaf out of their playbook and acknowledging the need for community politics. Far-right groups across Europe have used similar locally embedded strategies with scarily similar success.

Parliamentary and institutional politics are not the only avenues. Community power matters too. But they cannot be thought of as mutually exclusive. Electoral politics, we know, is about more than voting and you can't win the game if you don't play. Many progress-ives seem constitutionally allergic to party politics. This is letting the perfect be the enemy of the good and, no, that is not a sure-fire route to compromising political integrity. It just takes a bit of skill. Unfortunately, the Right are ordinarily better at it.[53]

In the UK, there has been a swing back towards eco-socialist and progressive politics in local government, at least in some areas. In 2021 the Green Party more than doubled their council seats, taking votes from all three major parties.[54] The Welsh government is run by socialist Labour members and they are introducing a raft of ambitious proposals, including rent controls. Labour councils across the rest of the nation have been instituting varied species of municipal socialism, from community wealth building and social housing to green new deal plans for local communities. Councillors Aneesa Akbar's and Matt White's website Building Socialism in Local Government offers training to people in local politics on progressive policies, as well as invitations to meet-ups. 'Out of power nationally,' they write, 'Labour can still make transformative change happen locally. Despite endless Tory cuts, Labour councils across the country can transform our communities.' Regardless of how you feel about Labour (or your national equivalent) we cannot leave the yawning gap between policy wonkery and grassroots action as an open goal for those who want to destroy everything we love.[55] (P.S. I found out whilst writing this that you can become a councillor even if you have been sectioned under the Mental Health Act. Who knew?)

Shared space and collective learning
Empathic and regenerative organising depends on shared physical space. Common space, along with the patient, careful and loving practices of 'the commons' have been carpet-bombed by those satiating the insatiable hunger for capital accumulation. The privatisation of land has progressively denied people the ability to support themselves, locking in a dependency on wage labour and simultaneously shrinking the spaces we are allowed to physically inhabit. But there are movements trying to change that.

Campaigns like Right to Roam, whose slogan is 'the right to roam is the right to reconnect' and Devon-based The Stars Are for Everyone, organise mass trespasses of land that used to be common property.[56] In April 2022 Kinder in Colour, a group of 400 Black people and people of colour, re-enacted the historic Kinder Scout mass trespass, the first time the anniversary of the 1932 event was celebrated by a BME organisation. They aimed to shine a light on 'the history of the countryside's rootedness in colonialism and exclusion'.[57] According to their figures, only 1 per cent of UK ethnic minorities access the countryside.

In cities, too, it is hard to find anywhere to be with people where you don't have to pay for the privilege, other than your own 'private' (usually rented) space. Civic Square in Birmingham is a rare jewel of an exception.[58] They offer free public space in a beautiful square, invite people to engage in their 'neighbourhood lab' to investigate alternative economic and social paradigms like doughnut economics – all with an explicit focus on 'regenerative and social infrastructure'. Another part of their project is a neighbourhood platform called The Front Room. This is a shared space, physically represented by a pretty barge used as a floating community café on the canal. It is an effort to extend the traditional concept of the 'front room' into the neighbourhood, providing a space to meet, have a cup of tea and, they say: 'connect with others from the area and beyond, process the present and dream of the future together too.'[59] Reconnecting to community is key. We are hyper-connective animals, but the threads holding us together have been systematically cut away. We need spaces. If we have the means, or we know someone else who does, we should offer space to others. Free space to congregate is a necessary respite from the dehumanising, extractivist infrastructure we are forced to live in. It gives us a taste of what it could be like to be free.

There are other ways of congregating and building community through shared enjoyment. Suffragettes from New Zealand to the UK had women-only cycling clubs over one hundred years ago, as well as the tea rooms we have already talked about.[60] Throughout history and across the world, many of these activities have been educational tools – like the autonomous education programmes of Zapatista youth and the Pan African Saturday schools in the UK.[61] The Workers' Educational Association and Open University helped add fuel to the fervent reading culture of the British working class in the early 19th century. Today the Open University still exists, as does the Workers' Educational Association, albeit in diminished form, but we also have Unlearn University and crazily good content in the form of free e-books, edutainment YouTube videos and podcasts, as well as a number of radical reading groups that are still going strong. (You can find access to some of these in the further readings at the end of the book.)

There are Jacobin reading groups in Canada and the US, as well as a few more for the UK-based socialist magazine *Tribune*, including in Liverpool and Manchester, for instance. This doesn't replace action, it informs it. It connects you to others who might want to do it with you too. The Left Book Club goes one step further. Originally established in the UK in 1936, it was revived in 2015 with its slogan: 'Read. Debate. Organise.' Not only do they send out books, they also pull together events with authors and keep an up-to-date interactive map of all the open, decentralised reading groups you can attend.[62] Abolitionist Futures has reading groups on Zoom. They also offer audio versions of the texts (good for the blind and partially sighted, but also nice for people like me who would rather let some texts wash over us than scour every letter with our eyeballs).[63]

Sometimes I feel like learning to reconnect is like going back to nursery school. I don't mean that patronisingly. For me, it's refreshing. It takes me back to simple sayings like 'sharing is caring', which seem trite but, well, right. Regenerative activism, reconnecting from an intentionally naïve and humble base, is also about reconfiguring how we relate to other people and to the world at large on a basic level. It's about unlearning many of the encultured habits of capitalist domination, about playing around with new forms of being us. It is also about trying to feel comfortable with acting in a horizontal manner, being comfortable relinquishing power, making mistakes and learning and adapting as you go.

All of this can make us kinder to ourselves. It requires trust. Not trust in the system but rather trust in ourselves and those we are organising with. It is about finding a new and yet familiar home, a place in yourself that's secure, that you believe in and from which you can find, to butcher the last written words of Christopher McCandless, a happiness that is only real when shared.

10

REMEDY: how to change the things we cannot accept

'Soyez Realistes, Demandez L'Impossible (Be Realistic, Demand the Impossible)'

– May '68 slogan

'Un Mundo Quepan Muchos Mundos (A World Where Many Worlds Fit).'

– Zapatista slogan

In 2022 the UN announced that there was 'no credible pathway to 1.5 degrees Celsius' without what it called 'radical transformation'.[1] Of the 116 pathways laid out under the Paris Climate Accord, 101 require drawing carbon out of the atmosphere to reach net zero. In the whole world we still only have around twenty large commercially operating carbon-capture sites.[2] One IPCC 'overshoot' scenario puts the amount of land necessary to draw down carbon from the atmosphere at two-and-a-half times the size of India, growing monoculture forests then burning them and forcing the carbon underground. The 'limited overshoot' scenario still requires land roughly the size of India. This is land the planet does not have, and it is indigenous people and many others in the Global South who are already being evicted from their lands to accommodate 'geoengineering as dispossession'.[3]

On 18 January 2023 the *Guardian* released analysis that found 94 per cent of accredited rainforest carbon offsets were 'phantom credits'.[4] Fake, in other words. A few days later a tech entrepreneur illegally released balloons full of sulphur dioxide into the atmosphere above Mexico. He was beta-testing the efficacy of messing with solar radiation, trying to reflect more of the sun's energy back into space using an atmospheric hack that any real-life Bond villain could unilaterally achieve (along with all the acid rain and ripple effects that filling the air with sulphur could bring about).[5]

These are some of the death throes of the neoliberal capitalist class as they struggle to accept the manifest injustices that extractivism has wrought on people and the planet. The proposal that we all drive electric cars is a prime instance of this epistemological dementia. Car culture simply cannot continue. Manufacturing just the batteries for electric cars to replace the existing car fleet worldwide could gobble up 6.5 per cent of our remaining carbon budget.[6] Many experts claim we don't have enough lithium to replace existing infrastructure with renewables; mainly cars and grid storage.[78] Most lithium reserves are on indigenous and Global South territory.[9] *Foreign Policy* warns that the scramble for critical minerals threatens a new 'Cold War' between China and the US.[10,11] In the space of a year, lithium's price has increased ten-fold.[12] Lithium also happens to be humanity's oldest psychiatric drug with a documented history in pre-industrial cultures all over the world.[13] I take lithium every night. I need it, as do millions of others. I'd really prefer it didn't run out.[14]

We inherited a culture of disconnection and domination. That is what is driving this. We need reconnection and equity, for our planet and for our minds. What we need is radical transformation.

Green New Deal

A Green New Deal (GND)[15] hinges on using a huge injection of government funds to kickstart decarbonisation. Its name comes from US President Franklin D. Roosevelt's New Deal of the 1930s, which used then new Keynesian economic theory to directly combat huge unemployment levels and a historic economic crash. The New Deal used unprecedented sums of government money to create new jobs, massive infrastructure projects and nature restoration, as well as funding the arts, education, mental-health support and direct cash assistance to the poor. It introduced the country's first social safety net, including minimum and maximum wages in some sectors, disability insurance and state benefits for the unemployed. It saved hundreds of millions of people from the scourge of a period literally named the Great Depression. Unsurprisingly, this level of intervention had an incredibly positive impact on people's mental health, on top of everything else.

Today, GND could mean any country using public funds to build huge renewable-energy projects and create jobs insulating homes, building flood defences and emergency response systems. These are important interventions, but ones that sit comfortably within the schema of the social and economic system that got us here. Others, like those proposed in *Perspectives on a Green New Deal,* demonstrate how we need to have a qualitative shift, to use a GND as a way of changing the system itself. It is a chance not to greenwash, but to collectively redesign society so we can put wellbeing and ecological symbiosis at the centre, rather than exponential GDP growth, extractive and oppressive production processes and vastly unequal distributions of wealth and income. A neoliberal capitalist government and a radical eco-socialist ruling party could both introduce versions of a 'Green New Deal', without a hint of irony.

The Green New Deal I want to fight for is one that does not simply reduce our emissions, but that deepens our democracy and transforms the economic system so that everyone, at the very least, has access to the basics they need to thrive. This GND would be guided by principles of climate justice. It would also be a joyous pursuit, one in which everyone could engage. The 'world in which many worlds fit' – the Zapatista slogan quoted at the head of this chapter – will be achieved not just through sombre, serious, dedicated activism but also by enlivening communities to express their own cultures and by practising collective joy. We need carnivals, festivals, massive art projects. We don't know what the end goal of systemic transformation will be and celebrating together is a key way of figuring that out.

US Congressperson Alexandria Ocasio Cortez (AOC) painted a version of this in a video she made with Naomi Klein and *The Intercept*, illustrated by Molly Crabapple. In a beautiful example of future history, AOC says: 'We didn't just change the infrastructure. We changed how we did things. We became a society that was not only modern and wealthy but dignified and humane too.'[16] AOC has been advocating for increases to the minimum wage as part of a GND, as well as universal healthcare, versions of universal basic income and, drawing inspiration from FDR's original New Deal programmes, the revivifying of the Nature Conservancy Corps (NCC). The NCC offered 1930s-era Americans a guaranteed job restoring the environment. Today's version would see rewilding as a core part of that equation, as well as reconnecting to nature proper, along with all the mental-health benefits that provides. A transformative GND would create millions of jobs, perhaps even a jobs guarantee. It would be a just transition, supporting everyone to transition to a new way of life. It would let us prioritise the things we care about, including low-carbon jobs like social care, education and mental-health support, alongside low-carbon activities like play, creativity and rest. The focus of our economies would shift. Gross Domestic Product would matter far less than whether we have the essentials in place.[17] A Green New Deal would radically reduce inequality and poverty, whilst redistributing political power and agency to the many.

How, then, do we bridge the chasm between us as individuals and a 'real utopia'[18] version of a Green New Deal? How do we practise what Manda Scott refers to as 'thrutopias'? The Left is very good at imagining macro-level proposals, but it is sometimes difficult to see where we fit in, other than as passive observers of a battle it looks like we're losing. That is a failure of storytelling and strategy. Somehow, by default, we feel we are playing the neoliberal capitalist game – winning or losing according to its rubric. We feel we are complicit in it – or at the very least that it is the setting within which we are forced to operate. In fighting for a better world we are struggling to believe another world is possible, which is a similar dynamic to being mad. Why is it that there is a binary? Why must the 'other world(s)' be non-existent right up until the point at which there is enough momentum for it/them to feel inevitable?

The principles and values that will make up those alternative futures are with us already. So are many of the processes and relationships. As a patient in a psychiatric unit we are forced to pick a side between reality and madness. So are activists. So is anyone who believes things can, should and will be different. But it is not black and white. 'Reality' is 'real', in so far as it is one perspective on the universe. 'Madness' is just as 'real'. There are as many realities at any one time as there are conscious beings.

In the context of a world so insistent on telling us to follow a strict path, we also need to have confidence in ourselves and our own truths, bolstered by those of the people we love. A man adept in this was Thomas Sankara – the revolutionary first President of Burkina Faso. As he said:

> You cannot carry out fundamental change without a certain amount of madness. In this case, it comes from nonconformity, the courage to turn your back on the old formulas, the courage to invent the future. Besides, it took the madmen of yesterday for us to be able to act with extreme clarity today. I want to be one of those madmen.

To cast the net even further back in time, I like to use Freud's definition of 'madman': 'The Madman is a dreamer, awake.'[19]

We need to band together and get on with it. We can start by identifying bridges in the space between the individual and the system. When I spoke to Dr Emma Lawrance, the author of several prominent papers on eco-anxiety, based at Imperial College London, she framed it as the need to 'connect the individual and the community level, the community level with the system, all in a way that they work together'. She added: 'People often talk about action being the cure for eco-anxiety, but it has to be the right kind of action and, crucially, people need the right kind of support.' I agree.

Ruth Potts helps to co-ordinate the Global Alliance for a Green New Deal. This is an alliance of twenty-seven lawmakers from twenty-two countries dedicated to implementing ambitious, transformative programmes. Ruth was also a member of the team at the New Economics Foundation when the organisation coined the phrase 'Green New Deal' back in 2008. When I spoke to her about how people can push forward a GND, we initially discussed their joining groups like the SunRise Movement in the US and GND Rising in the UK, both valiant youth-led campaigns to push forward the agenda by confronting politicians and getting impressive amounts of media coverage. But these are not the only options. 'There's no one action, no one way of organising that will solve things,' she said. 'All of us need to find a place that nourishes us so that we don't burn out.'

Ruth knows a lot about burnout. Amongst (many) other things she was one of the Stansted 15, a group of activists convicted of terror-related charges after blocking a plane that was set to deport sixty people to West Africa in 2017. The intersectional coalition of fifteen activists – from the organisations End Deportations, Lesbians and Gays Support the Migrants, and Plane Stupid – lived with the convictions until January 2021 when the Court of Appeal overturned them on the basis that there had been 'no case to answer'. As Ruth tells me:

We need to find a way of actively engaging with the world, with the enormity of the crisis, by finding our place in the process of shift – having some change as part of a series, in the hope that that will lead to wider transformation.

I have a lot of time for stuff that is more collective. There's a big difference between going for a walk in the woods or practising self-sufficiency, and going to a community garden, getting involved in mutual-aid networks or potluck suppers. One cultivates stillness. The other cultivates interconnection and an understanding of responsibility in and for the world – a positive thing, rather than a burden. You do need places of stillness, but if we can't cultivate community we're going nowhere.

As we saw in the previous chapter, community can be a wellspring of action. Actions rooted in community bring connectedness, belonging and purpose in ways that other kinds of action struggle to achieve. Connecting to national organisations and their local chapters is one place to start. In the UK, that includes Transition Together/Transition Towns, Green New Deal Rising, Stop Cambo, Labour for a Green New Deal, The World Transformed and Scottish Communities Climate Action Network, in addition to hundreds of locally based mutual-aid networks.

But wherever you are in the world, there are already lots of resources online about how to DIY retrofit homes, to use domestic energy efficiency as just one example that addresses climate change, mental health and economic justice together. The Community Energy Project, based in Portland, Oregon, has a wealth of free online material, including a year-round DIY weatherisation workshop. They also offer free physical materials to help with installation (on application) to those based in the area. Learning how to do this yourself could then be a launchpad for running free training workshops in your community and offering to volunteer to help your neighbours adapt their homes. You could start by connecting with others in your area, skill up, then expand it into the community as a mutual-aid group. With the right collaboration and planning, that could spin out into a campaign to demonstrate to local and national government how easy it is to get this done. If we can do it, why can't you?

Ultimately, we are hoping for thick networks of alternative ways of living that are enjoyable to be part of. We need collaboration between community organising, campaigning and more established organisations. These are boundaries that themselves often blur. Leo Murray, for instance, has been involved in implementing a local cycle highway and low traffic neighbourhood, campaigned with organisations 10:10 and Plane Stupid (as part of the latter he climbed onto the roofs of the UK Houses of Parliament and threw paper aeroplanes from there). He is also the Director of Innovation at the climate charity Possible. Possible's campaigns often have a distinctly participatory element. Their website has a good (and quite internationally transferable) guide on how to build local networks and get to know your neighbours. One of their energy projects, 'powering parks', worked with Hackney council in London to install heat pumps in open spaces. They have a toolkit and contact details so members of the public can take up the idea and run with it (Possible calculated that London has enough green space to heat 5 million homes). They also recently opened a branch of The Fixing Factory in Queen's Crescent, north London, as a community space for people to get help mending their stuff rather than throwing it away. In a related initiative, campaigners in New York won a campaign to force companies to provide instructions on fixing broken products. This is the first state-wide 'Fair Repair Act' implemented in the United States, the nation driving our culture of consumerism, extractivism and disposability. It is being fought in the courts and has suffered from watering down but it's more than a crack in the wall. A new economic system as part of a Green New Deal will require as many of us as possible, working on many different levels, to invent the future as we go.

Process is part of the puzzle. When I asked Leo whether he thought the work he does was good for him, he told me: 'Since I first confronted a now familiar overwhelming sense of existential dread about the accelerating environmental collapse, around seventeen years ago, I've tried many different coping strategies to keep it together. The only thing that has been consistently effective at keeping these feelings at bay is action – the more challenging and wholehearted the better.'

Mutual aid
It took a global pandemic to bring mutual-aid groups into the limelight. They were vital in keeping many people alive, providing food, funds, equipment, transport, conversation (socially distanced

or virtual), company, entertainment and deliveries to millions if not billions. Mutual-aid groups provided lifelines or, more accurately, 'life-webs' – networks of collaboration where everyone played a role. This was in stark contrast to the conventional uni-directional charity approach, soaked as it often is in a potentially patronising and disempowering saviour dynamic. People were isolated in their homes, out of work and divorced from all their normal social ties. In many countries, mutual-aid groups came to the rescue.

We saw 4,000 new groups established in the UK alone. More than 1,600 continued operating after the nation went 'back to normal'.[20] They continue to meet the needs of thousands of people with little to no public funding.[21] These, and thousands of other mutual-aid groups around the world, help people address immediate material and emotional survival needs, and build a shared analysis of the systemic causes of suffering. They persist largely due to the positive mental-health impacts of active solidarity. They are key to building strong and durable movements.

In 1902 Peter Kropotkin, the Russian anarchist-socialist prince, argued that altruistic, mutualist behaviour was a more powerful driver of successful evolution than selfishness and competition. His book, *Mutual Aid*, caused a storm. Rooted in evidence-based science, it was a strong rebuke to Social Darwinism. Kropotkin deconstructed the notion of conventional economics' core theory of the individual as *homo economicus* – egotistical, pleasure-seeking agents in direct competition with others, pursuing infinite wants in a world of scarce resources. Kropotkin argues that we are far more than this. He wrote:

[We are] appealed to be guided in [our] acts, not merely by love, which is always personal, or at best tribal, but by the perception of [our] oneness with each human being. In the practice of mutual aid, which we can retrace to the earliest beginnings of evolution, we thus find the positive and undoubted origin of our ethical conceptions; and we can affirm that in the ethical progress of man[kind], mutual support not mutual struggle has had the leading part.[22]

Activist and author Dean Spade argues that modern applications of mutual aid are fundamental to systemic change. They help people collectively to discover their own ability for community-level self-sufficiency. Examples include the Black Panther Survival Programs of the 1960s, with their free children's breakfast clubs and rides for the elderly, the Young Lords in Puerto Rico delivering medical services and eviction support, Socialist Feminists providing free emergency abortion services and gay and transgender community clinics. Movements that include mutual-aid programmes are inherently more popular, effective, sustainable and democratic.

Spade contrasts mutual-aid projects with more hierarchical charity and social-service programmes. Members of the former make decisions, rather than existing as either donors or recipients. Mutual-aid groups have open meetings, offer each other help unconditionally and embed the work in 'deep and wide principles of anti-capitalism, anti-imperialism, racial justice, gender justice and disability justice.' Spade's full taxonomy is available for free online.[23]

Some argue that mutual aid, as it is understood and practised today, could in fact act to prop up existing power structures of domination and dependency. Mutual aid, they argue, has been largely depoliticised, divorced from systemic analysis and horizontal education around power structures.[24] This can happen, but contemporary emergency mutual-aid groups have maintained deep connections with wider movement work. Renters' unions, food banks, communal eating programmes and hardship funds to support people struggling financially have all been officially and unofficially linked to survival programmes. Mutual aid is a form of community organising. Solidarity builds out into wider struggles, in contrast to much single-issue campaigning.

Mutual-aid groups can be hard to find independently, but the community-run online database MutualAid.wiki has a live map of almost 6,000 active groups.[25] This kind of organising is also key to building climate resilience and climate justice. As documented in books like Rebeccca Solnit's *A Paradise Built in Hell* and Naomi Klein's *How to Change Everything,* communities often spring up and provide each other with essential survival needs through effective, participatory and improvised mutual aid. Dr Elaine Flores, who is conducting a meta-analysis of climate-related disaster response, tells me: 'Community cohesion is the surest predictor of whether a community will cope, especially in terms of mental health.'

I was editing this whilst in New Zealand, when a cyclone hit. It ravaged much of North Island, particularly the Far North areas where a considerable portion of New Zealand's Māori citizens reside, with Te Hiku Iwi communities particularly affected. At one point 16,000 homes were without power. But communities that already had deep connections, like those served by the dozens of community organisations that gathered some months before the storm, thrived. They sprang into action when the cyclone hit, with stakeholders across the region meeting daily to co-ordinate support. Sarah Boniface, one of the Far North's only two emergency-management professionals, told the *New Zealand Herald* that 'you can be prepared as you like, but if there's no connection before it starts, it's slumpy'. The collaborations in the region, Boniface said, lead to 'epic resilience'. In the run-up to the cyclone New Zealanders were warned it was going to be the 'storm of the century'. A woman I spoke to in the aftermath said there 'seem to be one-hundred-year storms every few years now'. Getting to know each other and learning to work together is something we all need to practise, regardless of whatever is on its way. Connecting is also, thankfully, something we all want. Climate is a good excuse to start.

Deepening democracy, redistributing power
We live in a very stratified society. Even those of us who live in liberal democracies are fundamentally divorced from power. Much of what we have discussed so far relates to rebuilding power, from the community level to the movement level, but there are also opportunities for us to directly redress the injustices and imbalances of our exclusion from meaningful decision-making.

One example is organising Citizen's Assemblies. These have become increasingly popular in the last couple of decades, with radical and transformative iterations popping up around the world. These can be formally organised by the government – like Ireland's on abortion[26] – or independently set up by communities and organisations – like Iceland's following the Great Recession.[27] A random group of demographically representative individuals are selected to participate in the assembly, typically between 30 and 160 people. It is like a jury, only they collectively make policy recommendations and demands, having delved deeply into topics such as climate change, mental health, economic democracy, abortion rights, electoral reform and even the writing of constitutions. A Citizen's Assembly is crowd-sourced knowledge, collective wisdom and conflict resolution at its best. They call on experts, hold public hearings and are at the forefront of experimentation in deliberative, participatory democracy.

These experiences are starting to translate into the climate space. A Climate Assembly in the UK, commissioned by the House of Commons, came out with astonishingly progressive, complex and thoughtful recommendations for how to reach net zero by 2050.[28] The group's proposals were not binding, unfortunately, but were widely publicised and gave citizens an unprecedented chance to be directly involved in climate issues on a national level. There have now been Climate Assemblies in France, Hungary, Portugal, Poland, Ireland, Canada and the US.[29] The first Global Climate Assembly was held in the run-up to Glasgow's COP26. UN Secretary-General António Guterres, the man who called investing in new fossil fuels 'dangerous' and 'moral and economic madness,' described the Assembly as a much-needed practical demonstration of 'accelerat[ing] action through solidarity and people power.'[30] Climate Assemblies take a lot of work, but they are possible to set up independently and contributing to them or rallying support is easy. Extinction Rebellion has a website with resources on Climate Assemblies.[31] There is also a site sharing best practice on how to implement and run them, from Europe's Knowledge Network on Climate Assemblies.[32]

There is a long tradition of this kind of work in the Global South, but it is often omitted from the discussion, partly because the practices go by different names. South America has had its fair share of deliberative and participatory practices connecting millions of people. Movimento Sem Terra in Brazil has been running nested collectives for decades. The Plurinational State of Bolivia's 2009 constitution wrote participatory and communitarian democracy into law, especially for indigenous people.[33] It also established the central concept of Buen Vivir: good living and the right of nature as more important than economic growth (as had the Ecuadorian constitution of 2008).[34] More recently, Chile held a constitutional convention under the new progressive president, Gabriel Boric. It was a sprawling, hugely popular affair, building from local discussions up to regional and national deliberative processes. The proposed constitution that resulted was one of the most radically progressive in the world, explicitly discussed as an ecological constitution, emphasising a slew of rights for indigenous people and nature, as well as provisions for direct democracy. This was, unfortunately, voted down in September 2022.[35] It was not an all-out loss. The movement infrastructure and collective future-building exercise still carries much weight. It was a nationwide success in imagining real alternatives.

We cannot let democracy be reduced to periodic elections. Democracy Next argues that, at the very least, we need citizen representatives in government and new bodies like Citizens Assemblies to be made permanent institutions. Constitutional conventions should be part of the mix. So should an idea called sortition. Sortition is essentially government by lot. Like a jury, it is made up of randomly selected members of the public (balanced for demography), but it is more permanent. A branch of government selected by sortition would wield powers to propose, assess and amend laws – like a Citizen's Assembly, only with much more power. The UK Labour Party recently suggested that, if it wins power, it may dissolve the House of Lords. It's about time. I think it would be amazing to have a People's House in its place, a second chamber comprising members of the public, who could speak to the media about the processes of government, deliberate in the open and spread power out into communities around the nation.

Again, this might sound hard to sink your teeth into, but there are local projects all over the world enacting these kinds of principles. The influence multiplies. In Porto Alegre, in Brazil, a process of Participatory Budgeting (PB) has been running for decades – where the citizens have the power to decide how a significant portion of the local budget is spent in their community. The process in Porto Alegre has been partly co-opted, but the idea has spread widely. Now there are PB projects on every habitable continent.[36] There is no space to go into all the other accessible and engaging routes into democratic experimentation, like Community Wealth Building, Flat Pack Democracy and PARECON, but I will point to more resources at the end of the book.

Universal Basic Provision – UBI + UBS

Universal Basic Income is huge, both in its potential and its possibility for realisation. It is a radical social-assistance programme that involves giving residents regular direct cash transfers, universally and unconditionally. It has majority support in virtually every Global North nation polled, as well as a dedicated stable of politicians and organisations propelling it towards realisation. It is, as its advocates like to say: 'an idea whose time has come'. A basic income would drastically reduce inequality, economic precarity and poverty, increase the bargaining power of the 99 per cent, raise educational attainment and stimulate entrepreneurial activity, all whilst leading to no meaningful reduction in employment. A basic income could drastically improve health, too, especially mental health and general wellbeing.[37]

A basic income would have to be given solid guard rails. We would need rent controls, so that landlords couldn't just put up rents by the exact amount of the payment. There would have to be reforms to pay policy so that employers could not simply drop wages. It would need to be set in stone that other benefits, like those for disability and unemployment, would be maintained. In other words, we would all have to work to ensure that it was a meaningful transition away from economic exploitation. The exercise of figuring out what to retain, what to scrap and what to invent anew would be a daunting but emancipatory exercise in building a new system.

How is this linked to climate change, you may ask?

The biggest examples of actual basic-income implementation – in Alaska, Iran and Mongolia – are all funded by state-owned fossil fuels. What if we did the opposite: funded a basic income through a carbon tax? Prominent climate scientist James Hansen is one advocate of this plan, a so-called Carbon Fee and Dividend (CFD).[38] The Citizens' Climate Lobby, a grassroots organisation based in the US, are in favour too. They call it 'the policy that climate scientists and economists alike say is the fairest and most effective way of getting to zero carbon'.[39] A CFD has three main components.

First, there would be a flat tax on carbon, at the site of extraction and on all imports, meaning the price of goods and services would rise in direct proportion to how much carbon dioxide they have embedded in them. This would disincentivise high-carbon goods and favour more sustainable products.

Second, the cash raised would be redistributed to all residents, equally. By and large this would be overwhelmingly progressive. The poorest sections of the economy have the lowest emissions per capita, and the richest would end up putting much more into the communal pot.

This brings us to the third element – a series of safety mechanisms to make sure that the poorest do not lose out. There are going to be some people who have inefficient cars or draughty homes and businesses that rely on high energy inputs. It would be the government's responsibility to ensure that all of that is made as carbon efficient as possible, and to provide financial and technical support to those struggling financially because they are stuck in the fossil economy. As it stands, it is estimated that a CFD would financially benefit 95 per cent of low- and middle-income families in the US.[40]

Dealing with climate change is often framed as an economic threat. Researchers argue that a CFD would do the opposite:

'With progressive revenue recycling . . . aggressive climate action can pay large dividends for improving well-being, reducing inequality and alleviating poverty.'[41]

It might seem difficult to connect these huge ideas to personal action and the actual experience of mental-health issues. Jamie Cooke, head of the Royal Society of Arts International, had a brilliant suggestion:

> Community pressure can be really effective, particularly when it involves storytelling. People still have this *Benefits Street* mindset [a UK reality TV show shaming people reliant on government benefits], and if we can tell powerful, real stories about the incredible differences a small amount of unconditional money could make in people's lives then we'd be much closer to winning.

This could take the form of collecting stories about what people would do with the money and then presenting them to the council, a powerful tool for them to use in convincing relevant authorities to fund a pilot. Michael Tubbs, the first Black mayor of Stockton, California, set up a basic-income pilot off the back of similar storytelling projects. Tubbs, the youngest-ever mayor of a US city, had a personal drive too. He grew up with a young single mother and an incarcerated father. 'My belief in basic income,' he told CNBC, 'came from thinking how would my mom use that money.'[42] Other cities are following suit. There are now nearly fifty independent pilots across the nation.[43]

If the state is being recalcitrant, there are some good examples of charitable initiatives. GiveDirectly offers donated cash as a basic income to people living in Kibera, Kenya, the largest urban slum in Africa, as well as to Yemeni citizens, refugees in Uganda and poverty and disaster relief in the US.[44] This is charitable, yes. That means it is one-way and suffers from the risk of both patronising saviourism and individualistic giving. But one big advantage of a basic-income gift is that the money goes directly to people who need it – not as goods or services or anything else, but as cash. There is no intermediate process deciding what kinds of things people need, a process that can be deeply paternalistic. It gives people the ability to choose for themselves, based on their irrefutable expertise about their own lives and situations.

Sometimes I find it easier to think about this as radical redistribution. The end of 2022 saw the launch of the Hero Circle app. The project crowdfunds money to provide international climate-justice activists with a basic income. So many activists do this kind of work in their 'free' (read: unpaid) time. A basic income through an app can allow them to fully dedicate themselves to the struggle for climate justice.[45]

Some say Universal Basic Services are a better idea, providing free transport, energy, food, internet access and housing instead of cash. But basic income and basic services are often made into false competitors. They are complementary, and acting otherwise feeds into a dangerous scarcity mindset, one that constantly factionalises the Left. I think about UBI and UBS as part of a wider system of Universal Basic Provision. They both entail new money going to people universally. The differentiating factor is who gets to decide how that money is spent – individuals in UBI, the government in UBS. In different contexts, different ends of this decision-making scale are more appropriate. We can experiment, too. What would Universal Basic Provision (UBP) distributed by a community look like?

UBP, in its most radically progressive form, is about providing everyone with the basic necessities for thriving. There are ways to get directly engaged, build communities of support and stitch UBP together with climate justice and its huge benefits for mental health. It can be overwhelming to think about all of this systemically, but it doesn't rest on the shoulders of any one person. And it can be comforting to know there are millions of people pushing in the same direction as you.

Four-day week
The four-day week is another deceptively simple, but massive systemic intervention. The five-day working week is, many argue,

completely and utterly outdated. Unions won the weekend in the early 20th century. John Maynard Keynes, the most influential left-of-centre economist of the last century, thought that by now we would be working fifteen-hour weeks.[46] Our productivity has increased, but if anything we are working more (those who have jobs, anyway). Wages have stagnated in much of the world since the 2008 financial crisis. Overwork is one of the leading causes of depression, anxiety and suicide.[47]

Campaigners all over the world are arguing that it is time for us to cut our hours whilst suffering no reduction in pay. Trials across the world show that people's health improves, particularly their mental health, and overall productivity either stays the same or actually increases. There are major climate implications, too. One study claims that moving to a four-day week could reduce overall emissions by 20 per cent.[48] Gifting people a three-day weekend every week could also increase levels of community engagement, exploration of creative pursuits and, importantly, rest. As we know from Tricia Hersey, founder of The Nap Ministry: 'Rest is Resistance'.[49]

Time poverty is a real thing. Our time is finite on this planet and as much of it as possible should be ours to do with as we choose (individually, collectively). During the pandemic a huge number of people (four in five in the UK) 'gave back' to their communities. The same study, following the easing of restrictions, found that people were struggling to engage locally mainly because of, you guessed it, 'lack of time'.[50] A reduction in hours could be just the tonic the climate movement needs.

There is also strong evidence that a four-day week could increase gender equality, making it more likely that men take up their share of unpaid labour with their increased free time.[51]

If you run a company or know someone who does, look into the possibility of implementing a four-day week with no reduction in pay. Microsoft is trying it out, as are Shake Shack, KPMG, Deloitte and many smaller companies working in materials, construction, care and more. After a six-month trial in the UK, one hundred companies permanently moved thousands of staff onto a four-day week.[52] A similar shift is happening in the US. As Jon Leland, Vice President of Kickstarter and four-day-week champion put it to me: 'being able to give time back to people . . . is possibly the most valuable thing employers can do for their employees and the communities they live in.'

If you don't run a company, join the push. That could be on a national policy level, or just badgering your boss to try it in your workplace. Organisations like the 4 Day Week Campaign (UK), the think tank Autonomy and 4 Day Week Global each have plenty of materials to help you persuade whoever you need to.[53] Action for a Four Day Week even has a guide for employees who want to develop a trial in their workplace.[54] Alternatively, you can now look for a ready-made job with a four-day week. In the UK there is an online repository for all job openings accredited as four-day-week jobs, with an almost exact replica for jobs in the US, most of which are remote.[55] It is time to reclaim time.

Rewilding

In the past fifty years, the earth has lost 70 per cent of its mammal, bird, fish, reptile and amphibian populations.[56] Species are becoming extinct at a rate faster than at any time since the dinosaurs were wiped out 65 million years ago, and this has already been dubbed by many as the earth's 'sixth mass extinction'.[57] Our species accounts for 0.01 per cent of the total living mass on earth, but in the evolutionary blink of an eye we have been here we have altered 97 per cent of the earth's surface. The vast majority of this has occurred since the 18th century.[58] The fact that we are living in the Anthropocene is now undeniable.

Much of this can be reversed. Life on earth has a vibrant tendency to diversify, the planet fosters life in all its fractal and complex diversity. Most of the thriving ecosystems that remain (including 80 per cent of the world's biodiversity) are stewarded by indigenous people. But for much of the rest of the world we have irretrievably lost the human cultures and intricate land-expertise that aided our non-human kin.

Rewilding ecosystems is one method to partially reverse the trend towards species paucity. Called 'Rewilding,' the movement aims to speed up the regeneration of natural ecosystems by clearing waste, tending land, incorporating helpful trees and plants, like nurse plants, and reintroducing endemic animals that have been driven away by civilisation. There are now 120 documented Rewilding organisations around the world, with hundreds of projects.[59] We have already seen ecosystems 'rewilded' to a point where they can support the reintroduction of beavers, wild boar, red kite, toads, snakes and pine martens.[60] Rewilding Britain claims that dedicating 30 per cent of the land to this kind of regeneration could capture 12 per cent of the country's total emissions.[61] Rewilding Europe aims to cover half a million additional hectares by 2030.[62]

Rewilding is a potential system-changing practice. There are lots of ways to get involved, from volunteering with an organisation or starting a chapter to restore wild spaces (see guidelines in the footnotes if you are interested in this),[63] up to the dubious but sometimes valid and effective eco-tourism opportunities. Rewilding Britain has a twelve-step guide to joining in.[64] Rewilding is best applied when wedded to local indigenous relationships with the land, and much can be learned from land-based practices in much of the rest of the world, as well as in the history of the place you now live (which is statistically most likely to be a city). Increasing access to land, mass trespass, protest and direct action to protect wild nature and campaigning for changes to land use – such as revolutionising agricultural practices – will also be key in the ecosystem of tactics to restore the land and seas.

There are climate benefits and there are pronounced mental-health benefits, as well as profound lifestyle changes implied. But gifting land back to natural processes – replete with wildlife bridges and corridors for migration – is intrinsically something we need to do. On the grandest scale possible, how can we live on the most thriving version of the limited space we have, in the most compassionate and vivid ways?[65]

New economic futures

It is time to start putting grand alternatives in place around the world, alternatives rooted in reconnection and equity. One approach, mentioned above, is the Bolivian and Ecuadorian approach of *sumak kawsay* (in the original Quechua), or Buen Vivir (Spanish). Springing up from socialist-indigenous organisations in the Amazon in the 1990s, the approach centres on wellbeing and ecology as the main priorities of human activity. Similar approaches have been tried all over the world in efforts to dethrone *homo economicus* and the perennial threat of avarice and oppression. The Global South's Non-Aligned Movement, formed in the 1960s, not only made huge headway in decolonising much of the world, but also argued for rights for countries to develop their own political and economic systems without external intervention. They also wanted to end dependency through interdependency.[66] Their resolution was adopted by the UN in the 1970s, to little effect.[67] Brought back for a vote in 2022, much of the text was adopted.[68]

The modern climate-justice movement is mobilising a lot of similar energy, making demands for loss and damage payments, as well as reparations, free technology transfers, debt jubilees and a greater say for communities in the planning and implementation of climate mitigation and adaptation. The Global South, after all, is only responsible for 8 per cent of cumulative historical excess global emissions, despite being by far the most populated half of the globe.[69]

Others have pursued systemic change from a perspective of meta-structure: how the global economy functions and what, exactly, are humans aiming for anyway? Some have looked for alternatives to GDP, like Amartya Sen's Human Development Index and later work on wellbeing. The New Economics Foundation developed a Happy Planet Index and the nation of Bhutan has a measure of Gross National Happiness. These initiatives are imaginative and insightful, but often hard to implement effectively. In a similar attempt, Kate Raworth left her job at Oxfam to write a book: *Doughnut Economics*. The 'big vision',

Kate tells me, 'is to meet the means of all people within the needs of the living planet. The goal is not to grow – it is to thrive.'

Rather than rely on a proxy like GDP, *Doughnut Economics* maintains that we should focus more explicitly on what we want to achieve and then create more effective economic systems to deliver that. The 'Doughnut' itself is made up of an inner ring, a social foundation comprising our essentials for flourishing – nourishment, economic security, peace, political voice, education, health and so on. The outer ring is the ecological ceiling, consisting of what are known in earth system science as 'planetary boundaries', beyond which the earth's system cannot regulate itself. These include land-system change, chemical pollution, biodiversity loss, freshwater use and climate change.[70]

The Doughnut is a concept and tool now being used all over the world. The Doughnut Economics Action Lab (DEAL) works with thousands of individual members worldwide, from all walks of life as well as dozens of communities, cities and regions.[71] DEAL has an explicit climate and ecological focus, but when I asked Kate why mental health was not explicitly incorporated in the social dimension she was refreshingly up-front. 'The Doughnut is a response to economics, a reply to the mainstream, so it comes out of the framing of economics where mental health as a category hasn't appeared.' We talked about the social and ecological drivers of mental health and she said that many of the cities they are working with are now incorporating networks and community, and including both mental and physical wellbeing in their metrics of health. The Amsterdam Doughnut, for instance, included metrics for on anxiety and depression, tracing it back through the social fabric in search of reasons and solutions.

Because the social foundation of the Doughnut was developed off the back of the Sustainable Development Goals which, Kate tells me, 'tend to be specified in terms of individual rights', DEAL did not start with an analysis of communal rights or indigenous rights. They are moving in that direction, though. As Kate says: 'We humans largely thrive in connection with others, so now we ask "do you have someone to turn to?" As well as helping with economic security, that also supports and stabilises our mental health.' As we talked further, Kate stopped: 'There's that deep-dive question – what do we even mean by mental health? How do you measure it? Sometimes I lay the doughnut flat on its side and draw a spiral coming up out of it – turning it into a 3D doughnut. If we live inside the Doughnut, we can create that spiral: it's a spiral of wellbeing, belonging, creativity, a sense of security. I think mental health belongs there.'

People often find using the Doughnut empowering. It can offer a structure within which to start navigating a world of endless complexity, hostile systems and unpredictability. It is most often used collectively by a group. In Birmingham, Civic Square, for example, host community spaces that use the Doughnut as a 'platform to organise, whilst also encompassing a plurality of bold visions'.[72] After one international peer-to-peer doughnut session, Civic Square co-founder Imandeep Kaur described the experience as 'gorgeous'. She said: 'The peer journey taught me how much I need joyful, optimistic and open spaces for this work, [and] how much less tolerant I am of space[s] that aren't joyful, generative and hopeful now.'[73]

It is rare to find a framework that is useful, let alone joyous to use, and that fosters shared belonging. Yet some have sprung up around us in recent years. Three I have already mentioned in passing are the ongoing open-air experiments of libertarian municipalism and social ecology in the Kurdish territory of Rojava, the rebel Zapatista autonomous municipalities in Chiapas and the co-operative, associational settlements of Movimento Sem Terra in Brazil. The recent wave of abolitionist theory and practice could also be included. This is a systemic endeavour that includes but goes beyond prisons, police and criminalisation into collectively imagining alternative futures, establishing communities, institutions and economies of care. As Black Lives Matter co-founder Patrisse Cullors has put it, it is about: 'building out a different way of being, a way dedicated to the dignity of human beings'.[74]

Alternative imaginaries abound, but even if you're a fan of Solarpunk or Anarcho-Primitivism, Eco-Socialism or Libertarian

Municipalism, or Fully Automated Luxury Communism or Degrowth, two important things are often missing in the personal lived experience of our beliefs. First, practical outward-facing actions we can take in our lives to meaningfully engage with and build towards these grand ideas (at least in a way that feels enlivening). Second, many of us end up suffering from a lack of belonging – a pervasive, pessimistic alienation. We need bridging actions and we need communities we can belong to. They are sorely lacking today, but that doesn't mean we cannot build them.

I was recently at an event on the Green New Deal and degrowth. It was framed as a debate, but actually both speakers were hugely supportive of each other and shared much of the same analysis. At the end, the audience was asked to vote on whether they were in favour of a Green New Deal, Degrowth or both. The percentage who voted for both was in the high nineties. I later spoke to one of the speakers. Harpreet Kaur Paul is a human-rights lawyer and climate activist who co-edited the scintillating collection of essays *Perspectives on a Global Green New Deal* (available for free online).[75]

I told her that I couldn't understand why, in movement work, there can be so much acrimony and distrust, so much fracture, given the ability for us to have public conversations like the one she had just had. We are often so fractious and difficult with each other. She told me it has something to do with activists often: 'Putting personal trauma to the side when doing political work, despite the many triggers we have. We navigate that inter-personally, but it hasn't come out in our politics. There's something about the ecology of activism and the personal journeys in it that is being missed.'

Activism cannot be a thankless grind or a sacrificial duty. It has to be something transformative, for us and for our culture. Harpreet described the existing gap as 'existential'. I couldn't agree more, in every sense of the word. I asked her whether she thought we can learn to be different in movement spaces. She said: 'Our movements have to be kind and nurturing and compassionate and healing to attract the millions of people we need. It's not a question of whether we can or should, but how – because we have to.'

11

Movement culture: how to keep us safe and happy

'If . . . one avoids the linear, progressive, Time's- (killing)-arrow mode of the Techno-Heroic, and redefines technology and science as primarily cultural carrier bag rather than weapon of domination, one pleasant side effect is that science fiction can be seen as a far less rigid, narrow field, not necessarily Promethean or apocalyptic at all, and in fact less a mythological genre than a realistic one. It is a strange realism, but it is a strange reality.'

Ursula K. Le Guin, *Carrier Bag Theory of Fiction*, 1986

EVERY THEME THIS BOOK has explored has its complexities. They are largely unknowable. The mind is complex. So is the climate. Organising social movements and inventing new social structures: that is a very complex business. We have a tendency to pretend we know much more than we do. Even with immensely complex computers and after centuries of exploration, the dominant culture cannot unravel these mysteries. We can sense tendencies, we can make predictions that often play out but really, we must learn to be comfortable with uncertainty. It is the defining feature of polycrisis.

In our hubris we have lost the ability to read the most important signals relevant to our survival. Complex systems announce when they are unstable but few of us can hear them anymore, let alone understand them fully. Complex systems are inherently more

intelligent than humans. They are older. Older beyond comprehension. There is survival logic in their branching diversity. This
survival logic is often interpreted as dysfunction. Like the mad,
these demands for transformation come from other planes, planes
the dominant culture has devalued and demonised. Climate
breakdown and mental breakdown are vital forms of feedback,
saying that something is deeply wrong and giving us coded
pointers for how to change – not to revert, but to transform.
The earth can use a flash flood to revivify and reform landscape.
A forest fire can bring forth new seeds. The psychotic mind can,
uncannily, scout ahead for meaning in uncertainty, an act of
exploration that has immense utility. The depressive mind can
strip reality back to a bleak but pragmatic objectivity. The anxious
mind can protect us from real dangers that others cannot see.
Dissociation can insulate us from terror, so that we can function.

All of this, experientially, can be deeply painful and extreme
both for ecologies and for minds. The hierarchical societies we
live in do not allow us to transform these experiences into widespread change. But, to an extent, we do get to choose the ways
our wounds heal and how we interpret the characterful patterns
they leave on our bodies. Together, without any kind of permis-
sion, we can choose to embrace a more horizontal, intersubjective
way of being. There is wisdom and resilience in open, dynamic
groups in the same way there is wisdom and resilience in diverse
ecologies. To survive and thrive, we have to build on a wide
range of relationships with each other, drawing on the creativity,
care and security we will need for the coming era.

Being comfortable whilst in a state of uncertainty – the defining
condition of polycrisis – does not mean being docile and placid.
If we are resilient enough, uncertainty can be a spur to action
because it suggests that things can change. The climate, the mind,
organising: these are all ecosystems of their own, continuously
interacting with each other. They are fluid, changeable, malleable.
We are already actively participating whether we like it or not.

In the process of researching, writing and interviewing amazing
people for this book, I find myself coming to the end of it more

confused than when I started. I am pleasantly confused, but not rudderless – quite the opposite. The more complex and uncertain our futures become, the less grip the dominant authorities have on any monopoly of truth and agency. The more complex and uncertain things become, the more issues we realise we are connected to, the more existing struggles we align with and the more people's lives we impact.

Climate change is just one of many factors that skew our mental landscape. It is an important one that links to many others – poverty, inequality, racialisation, gender discrimination, exploitation, disconnection, domination – but it is just one of many. By now we know that all the others are, at their root, the same. Individually, they are their own morbid branches, intertwined but distinct. Each branch bears its own sorrowful fruit, bitter and poisonous. They all emanate from the same sick tree.

Extractivism is at the root

This tree's roots are wrapped around a malignant mass: extractivism. Extractivism forces the insatiable pursuit of one thing: extracting life from living things. It sucks the life from lands, from bodies, from minds. This driving force, the engine within civilisational expansion, has been rebranded many times. It has been called Provenance, Progress, Empire and even Enlightenment. It has been called Capitalism, Industry, Neoliberalism and Growth. Its next iteration will likely be technological. It is all the same thing: the relentless transformation of the living into the dead. But with each iteration it is speeding up and spreading out. The ramifications for the human and ecological psyche are dire.

Extractivism has two beating hearts, the first of which is disconnection. By disconnecting us from our landbases and from each other (whichever came first in the seas of time), the second beating heart was allowed to come into being: domination. It was only by putting psychological, emotional and spiritual separation between each other as humans, and between ourselves and the earth,[1] by disconnecting us, that a species as ultrasocial and collaborative as ours could tip over into a mechanistic dynamic of oppressor and oppressed.[2] Human and non-human lives were then disposable. Social systems multiplied these inequalities of power and those at the zenith prospered.

Landing here was never a materially or socially pre-determined outcome. For thousands of years, vibrant and diverse cultures have flowed in and out of different social arrangements, right up until the point where stratified societies which had whipped 'man' and 'nature' into submission forced everybody else to live by their ways.[3] Today, extractivism is almost ubiquitous. There are hold-outs and pockets of different cultures, but just as extractivism led to climate change and there is no planet B, there is no-one alive (so far as I know) who is completely beyond extractivism's coercive reach. Climate pain, as Abi Deivanayagam writes, 'is salt poured on our existing wounds.'[4]

Fighting extractivism at the root

It might sound scary to suggest that we make the problem bigger for ourselves, but reimagining and building the world anew – whether that means our inner worlds, our social realities or the climate – requires a willingness to unlearn some things and explore other truths. And that is a scary thing to do.

I spoke with Rhiannon Osborne about systemic activism in this context. She told me she was deeply depressed as a medical student, in part because she felt she was being trained to apply sticking-plaster solutions to deep injustices. She saw children with asthma discharged to damp, mouldy homes, and homeless patients with recurrent infections unable to heal without a home. It was not until she found a political analysis of health and an organising community of mutual care and radical imagination that she fully recovered. Rhiannon is now a climate health-justice organiser, working with groups like The People's Health Movement and Ken Henshaw's We The People in the Niger Delta.

The individualisation of health, including mental health, is a tool of disguise to hide the health injustice of the extractive economy, the embodied stress of an exploitative employer

231

or the mental-health impacts of pollution. It's also designed to further fuel extractive industries, because if your illness is because of your behaviour or genetics, then it's also your responsibility to buy the right app, to sign up to a gym membership. You have to do something about this, and that can be consumer related. It's both hiding the violence of extractivism and encouraging further extractivism. Alternative visions which see health as collective, political and ecological can be a powerful tool for organising and healing and a deep threat to systems which are designed to create sickness.

Araceli Camargo agrees. Araceli is a neuroscientist of Turtle Island descent who founded Centric Lab – a neuroscience lab creating radical infrastructure for community healing.

When one is seeking to solve a problem, one must identify the core inputs that contribute to the problem rather than focusing on the outputs of the problem. Climate change is the output or the consequence of planetary dysregulation. Therefore it is essential that we learn the factors that contribute to the dysregulation of our planetary systems. One of those factors is the contamination of Soil, Air, and Water. When these Life creators are contaminated, it changes the microbiome (the micro-organisms and building blocks of Air, Water, and Soil). This in turn makes changes to biodiversity, including the death of ecological systems. When forests or wetlands die off it creates mass changes to local-ised weather and with time this leads to the climatic changes we are witnessing. The goal is not simply climate adaptation, the goal is to reframe, restructure and heal our relationship with Nature. Our health is tied to planetary health. Nature is our partner; therefore we must turn to them for guidance and leadership rather than imposing a top–down approach.[5]

When I spoke to the vacuum cleaner (James Leadbitter), an artist and mental-health activist, he made a similar series of points.

For James, a proudly 'mad' disruptor and organiser, addressing the problem at the root is vital, but not enough. He has been equipping young people in UK mental-health hospitals with the skills to organise for radically better mental-health care – a project called Balmy Army.

> That is of therapeutic value to them. But it's useless if they can't also process childhood trauma, have access to therapy and relationships, access to medication (if that is of benefit to them), and also housing, and benefits, and food, and not being subject to knife violence. So, organising isn't a magic bullet because you have to look at systems of violence and oppression and how they link to mental health, but young people organising to improve the material conditions of their healthcare can be incredible.

Looking after each other along the way

To fight extractivism at the root, we have to look after each other along the way. We are also best positioned if we ally with those who know more about these fights: how to survive them, and how to thrive. The climate movement is a relatively young one, and it is not always a healing place. The movement's culture is infused with exploitation, with power-hungry machismo, competitive brinkmanship, showboating and the glamorisation of status. There is very little space for genuine stillness, for healing and for joy. There is also little space for meaningfully participating in decision-making, for shared agency and mutual care: such spaces do exist, but they are all too rare. The corporatisation and callousness of lots of activism is grim and is making the sick sicker.

The founder of Healing Justice LDN, Farzana Khan, talks about this well.

> A lot of us are coming to movement work because we're so traumatised, because we've been hurt, because we've been injured, and then we reproduce a lot of those dynamics. Often, they're not from ill intention, but they cut. So how

do we make our movements robust enough, sustainable enough? How do we get skilful in order to become . . . an ecosystem? Somatics has increased my capacity for change, but it's also increased my capacity for growth. We need to mass mobilise right now, like: shit is real. We can't stay small. We can't stay afraid. We need to organise with everyone that wants to organise.'[6]

There are deeply rooted wisdoms in every culture around the world about how to resist, reconnect and remedy. There are more experienced movements, with modern iterations, who can show us how it's done. Some of the cases I will look at aren't directly connected to climate, but are relevant. Others are rare examples of healing collective space within the climate-justice movement. These movements value:

- Shared power
- Radical inclusivity
- Mutual care
- Embracing uncertainty
- Experimentation
- Embodiment; and
- Visioning

Learning from other movements
The contemporary feminist movement in Latin America is as much a culture-shift as it is a movement. They have successfully brought bodily and reproductive autonomy, domestic violence and gender inequalities at work into public discourse. Latin American feminists have not held these up as isolated issues but have rather trailed a thread back to colonialism and neoliberalism, the systemic oppression of the most vulnerable and pointed to the roles people have been forced to inhabit by structural forces. They also reject conventional organising models, having built a movement culture that is generous, supportive and loving. The principle of mutual care is a priority.

Practices have grown up in the movement that allow for a radical inclusivity. Many similar movements are only accessible to those who are relatively time rich, economically secure, able-bodied and 'able-minded' with no dependents. Here, however, mothers with young children are often offered childcare so they can join meetings. This means people from more backgrounds can share their knowledge. It also makes the movement more durable, bonding its members together and bonding them to the movement. A culture of care is strategic, as well as just.

Similar models are embraced in Zapatista autonomous spaces within Mexico, in the Kurdish region in Rojava, by MST (the landless workers' movement) in Brazil and the sprawling networks of co-operatives in Kerala, India. 'Groups like these are built for the long run,' says Patricio Provencio, a Mexican climate organiser and community educator.

They ground their struggles in the needs of the community. To do that, you have to be truly democratic. Local communities know best what they need. If you impose your vision onto an oppressed community, it's a moral mistake and it's not a resilient structure. There are material gains to centring communities of care in movement work, like growing group sizes, campaign wins, mutual aid and improved material circumstances. There are psychological gains, too. Everyone's looking for connection. Everyone can relate to talking about mental health. It can draw people in, and open things up. I've seen men open up in these spaces in ways they've never done before. The same can happen between generations. Just opening the movement up, sharing power and introducing mutual care, that can be a community education tool. We learn democracy together just like we learn mental-health resilience together.

Disability-justice groups are really good at mutual care, as those self-organising at the sharp end of oppression often are by design. In the UK in the early 1970s there was even an improvised, effective Mental Patients' Union. A stone's throw from where I grew up, a small group of patients at Paddington Day Hospital announced the formation of an association by and for the 'mental' to resist oppressive psychiatry, arguing that it was 'a form of social control of the working classes in a capitalist state, and that the psychiatrist was the "high priest" of technological society, exorcising the "devils" of social distress.'[7] The Mental Patients' Union published an influential pamphlet known as the Fish Pamphlet. The cover is a fish struggling on a hook. The fish represented the mentally ill, signifying that to the outside observer a fish on a line might look mad in its flapping, but really it is a natural response to a barb that threatens it with death.[8]

Today, organisations like Disabled People Against the Cuts in the UK, the West Coast Prison Justice Society in Canada and Istanà KSJ in Indonesia continue the tradition of breaking chains and making change whilst practising democracy, radical inclusivity and community reciprocity. STOPSIM is a campaign in the UK fighting recent moves to incarcerate and criminalise mental-health patients. STOPSIM have a buddy system. Much like many Alcoholics Anonymous and other 12-step groups (controversial but effective old-school mental-health mutual-aid movements) if someone can't make a meeting or an action, there is a 'second' to step in and take the reins and others to check in. This makes it accessible and allows things to run smoothly. If a movement isn't radically accessible, what the hell are you doing? If it's not accessible to people using wheelchairs or people hearing voices, it's not really a movement. This is not defensive action or a concession on the part of able-bodied, 'able-minded' people, it's strategic. There is strength in diversity and people on the frontlines need to be not just heard, but allowed to lead.

Globally, disability justice differs greatly. In Ghana there are disabled communities running community-education projects to try to end the recent practice of mentally ill people being tied to trees. There are similar initiatives in rural Nigeria, where the work of groups like the Mentally Aware Nigeria Initiative and

Jennifer Uchendu's The Eco-anxiety in Africa Project, both run by and for people with mental-health issues, is already radically altering discourse and lived experience. We need all of this.

As Jocelyn Longdon (aka Climate in Colour) points out, in Ghana there's a saying '*dua koro gye mframa a ebu*' ('an isolated tree cannot stand the might of strong wind or storm'), whilst in Nigeria it's '*Ig ikan kò leè dá gbó se*' ('a tree cannot make a forest').[9] Globally there are 1.7 billion people with disabilities. Together, we make up the world's third biggest economic power, behind only China and the US.[10]

James Leadbitter fights for Mad Pride, a movement that grew out of psychiatric survivor discourse. James was the first person to introduce me to the idea that a breakdown could be a break-through. He has worked with the Wellcome Collection, NHS and others to collaboratively design and build Madlove, a 'safe space to go mad'. He has also just finished a film, *They Let In The Light,* written by young people in a mental-health hospital and performed by the hospital staff. The process is all about breaking madness open, revealing its incandescent beauty as well as its devastation. Its strangeness. The reality in unreality. To do that justice, you have to respect the process. James, who has a long and winding background in climate activism and works internationally, tells me: 'There are tonnes of strategies that won't have been formalised and others that won't be findable because of how white-centred and extractive everything is.' One thing is for sure: 'If you have bad process, you get bad outcomes. If you have good process, you get good outcomes.'

Leah Lakshmi Piepzna-Samarasinha paints vivid portraits from the inside of the burgeoning disability-justice movement on the US West Coast. In *Care Work: Dreaming Disability Justice,* she insists we can build a better world in the here and now. 'We can do things differently, as organizers,' she writes. 'We do it all the time.' Instead of the 'typical movement burnout way – speed, panic, conference calls, panic, no food or breath or sleep,' she suggests raising funds for each other's support, physical exercises like somatic grounding to help safely process trauma together, cooking food, praying (if you do) and sharing things like food and rides.

> This is the kind of movement I want to be part of. I want movements to embody a disabled, working-class, brown sustainability that celebrates femme organizer genius. We deserve nothing less. And we – disabled, working-class femmes of color – have been creating these kinds of movements for a long time. Listen up. (Or read the captioning.)

Leah Lakshmi is also a climate-justice organiser. And a Covid-justice organiser. And a Trans-justice organiser. The movement work she does rejects the white-centred single-issue approach. It respects uncertainty, as all realistic work of this kind must. It respects the sprawl and interconnection of systemic oppressions. It respects the needs of the movement's members, including the need for healing, rest and joy.

Healing Justice

Healing Justice is a gift offered to all of us, a health-focused form of transformative justice with roots in Black liberation. It is a theory, but also a practice. It is used by movements, including Black Lives Matter, as an element of what abolitionists describe as transformative justice.[11] Healing Justice is a way of supporting individuals, communities and movements that is rooted in embodying and processing trauma. It tackles the systemic roots of historical and ongoing strife. It's not just cerebral, it's physical. Healing Justice involves doing healing practices together, like bodywork, and learning practical ways to look after each other. It is a prefigurative politics, creating positive worlds through changes in how we think, communicate, relate to each other and organise together. It is both a healing practice and a rigorous change-making strategy.

Somatic experiencing (somatics for short) is a key part of this: a set of inclusive, integrated regenerative techniques for mind and body. Somatics was institutionalised in the Global North by a student of Sigmund Freud's, Wilhelm Reich. Reich refused to accept that psychological issues could be confined to the psyche and somatics was his attempt to combine physical and mental healing. Its practice today has far wider roots and off shoots in cultures all around the world. Politicised somatics and generative somatics are relatively young disciplines, combining power analysis and strategy with this older psychological and physiological work.[12]

In *The Politics of Trauma*, Staci K. Haines, founder of Generative Somatics, explains the importance of listening to the body and sensation rather than thinking our way out of mental difficulties. 'The body/soma is the ground for transformation. We change ourselves through the body – and the body will ask more from us than the mind, in the transformative process.'[13] For Haines, processing trauma has to be part of a wholesale trans-formation, one that involves letting complicated feelings arise, changing how we react to difficult sensations and rooting our actions explicitly in our values and relationships – to the extent that conflict can become a site of 'generative' healing. This involves both individual work and collective work – doing healing practices together and building new forms of relationship. As Haines' emphasis on the systemic causes of trauma makes clear, the natural conclusion is to use recovery as a way of changing the world, and changing the world as a way of recovering. This is a never-ending process of building what Staines calls 'new shapes'.

When I asked interviewees based in the UK to give me their favourite examples of movement culture, Healing Justice London (HJL) came up every time. HJL has made a huge impact in a relatively short period of time. I spoke to their National Movement Co-ordinator, Ewa Jasiewicz. Honestly, it was a joy speaking to her after hearing so many people I deeply respect speak so highly of HJL's work. I asked Ewa to describe what it was that made HJL so distinct.

HJL really does start with the body, with those who have been deeply marginalised, whose bodies were not meant to survive, or exist, or have access to the commons. We're enabling as much community and as much undoing of

colonial, extractive, Eurocentric legacies as we can, legacies
that everyone has been shaped by . . . I see HJL as a radical
public-health practice, and a portal to the kinds of
communities and processes that we need to transform the
world. . . . We need healthy ways of relating to and organising ourselves toward each other and to find ways to
co-regulate a kind of transformative resilience. We don't
want to regulate and adapt to oppression, we want to resist
and transform it . . . [HJL is a] very quiet, capacity and
infrastructure designing and building movement. A sort of
fugitive movement. The people that [we] are working with
in HJL might not necessarily even identify as being part of
a movement. But they do identify with us, they do feel at
home and they do want to help grow the practice.

Now that pandemic restrictions have lifted in the UK, HJL have
been asking themselves what the most life-affirming infrastruc-
tures are that they can build. 'We're now on a trajectory towards
something much more hopeful.' HJL's long-term plan is to build
a health centre and apothecary in London. Over the next decade
or so they want to skill, train and support 150 movement leaders
in politicised somatics to enable greater capacity as community
first responders. News that there will be a growing, radical
healthcare movement in the city I live in made me feel much
more secure. It's something I hope I can be involved in, but it
is also a real comfort to know that the kinds of groups we need,
even physical institutions, are in the making. However the coming
decades unfold, we will need them, and they will need us.

More immediately, HJL are doing in-person events and
training sessions, whether in schools, GPs surgeries, organisations or unions where they can see more transformative justice
and process in action. There is a huge amount of excitement
here. Ewa recently wrote a piece on the collective healing potential of strike action – she has a decade's experience in trade-union
organising. Union people told her it was amazing and a totally
different perspective. 'That's because it's [a] new way to the

mainstream,' she tells me. 'This is what happens when ideas and practices that usually exist in separate bits of movements overcome their competition and start sharing.'

Urgency is not always helpful

Many people in the climate movement have an aversion to getting process right. Urgency is often the pillar they lean on for persuasive support. But urgency is often toxic in the climate space – both psychologically and tactically. My first major breakdown after the UN climate talks in Copenhagen followed months of repeating the phrase 'last chance to save the earth'. That was in 2009. Often urgency ends up meaning that the most excitedly or aggressively pushy get their way, performing learned habits of competitiveness and acquisitiveness in the process. This, like trauma snaking its way down through generations, results in movement cultures of impulsiveness, inequality and injustice. It is easy for that to be reproduced, no matter how well-meaning and sincere individual members are. After all, such behaviour arises from the only culture we have ever known. Abolitionist and scholar Ruth Wilson Gilmore talks about what she has called 'rehearsing freedoms':

> [We can] rehearse the social order coming into being, as against recite the complaints . . . rehearsing things like the possibility of retaking the streets as public space and then thinking about what does that mean – what does 'public' mean if not some ability to continually do and do again? What we are demanding in the short run, we need to be able to do for ourselves.

If we want another world, we need practice – in both senses of the word. We have been systematically deskilled in looking after each other – ourselves, society and ecology. We cannot pretend we have these skills if we don't. We cannot just make them up. We can educate ourselves, but truly the best way to learn is to embed ourselves in existing groups. By practising freedom.

The mad have tools

Seeing the connected canopy of extractivism's many branches and demanding radical inclusivity – especially through healing practices – opens up a space for the mad. Those of us whose minds meander are amongst those who find it most difficult to engage in movement spaces and can be at risk. But if we are held in the right community, the right ethic and the right principles, we have an extraordinary amount to offer and to gain. In *Care Work,* Leah Lakshmi Piepzna-Samarasinha elaborates on what she calls 'crip emotional intelligence' and 'crip skills' – a delicious subversion of Maslow's discrimatory and exclusionary psychology.[14] Using the terms 'meant something,' she writes, because 'the deficiency model by which most people view disability only sees disabled people as a lack, a defect, damaged goods, in need of cure. The idea that we have cultures, skills, science and technology runs counter to all of that. In a big way.'

I am not the first to suggest that the mad have idiosyncratic skills. The mad used to be shamans, oracles and accused of being witches. In all our splendid diversity, we tend to embody certain traits. Most of our behaviours are primarily gained through lived experience and the situations many of us are forced into by society. What follows is a list of what I would call some of our 'Mad Skills'. They won't apply to everyone, but they may be meaningful for some. Below this, there are some protective principles. Safety is a must. I struggle with this and it is going to be a lifelong learning process. The healthier I am, the more I want to do. I often overdo it, convincing myself beyond my malleable limits, then crash. It might sound boring – it still does to me – but the more I look after myself, the more I get to experience and engage. That is where the unexpected pays off. Not in the dramatic ruptures of my youth, but in the surprising revelations of propelling myself, alert, into the world.

Mad Skills

1. Empathy. Mental-health issues can teach us about the experience of suffering. Not all suffering is the same, but

living in a state of constant recovery tends to make it easier for most of us to put ourselves in another's shoes. That can help with connecting, with drive, with staying together. It can also make us pretty fluent in mutual care.

2. Patience. Anyone who has lived with mental-health issues knows it is largely a waiting game. We know the perils of mad-dash urgency.

3. Energy, surprises, zeal. Not every mad person is the stereo-typical creative wizard, but we do have a tendency to shake things up. We can occasionally skirt the outer edges of the universe or end up confined to our beds for horrifying stretches of time. When we are on our feet, it is easy for us to destroy the dominant culture's argument that 'this is just the way things are'. If we have confidence, we know from our own experience that other planes are plentiful.

4. Resilience. We mad have to bend with the wind without snapping. We can buck, we can buckle, we can have supports, but in order to survive we need techniques and practices, people and places that make us feel safe and allow us to live a life of malleable, adaptable, resilient recovery. Resilience must be a core value of any new system(s). That's our bread and butter.

5. Understanding the collective nature of health. Mental-health issues are disconnections from reality. We know, intuitively, that there is something wrong with the status quo, something that is making people sick. No wonder we often feel so disconnected from the 'normal' world. Finding other people who get that, too, is beautiful. Holding each other up is revelatory.

6. Priorities around rest and recuperation. Rest is not just responsible, it is revitalising. It is never 'doing nothing' (much as extractivist culture would have us believe). It is living. Rest is actively (and passively) loving our bodies and our minds. Movements need this wisdom.

7. Creativity. People with mental-health issues are forced to think laterally. We constantly have to navigate within and

around persecutory health systems, as well as work, relationships, medication and, for some of us, kaleidoscopic physical and psychological changes in perspective about ourselves and the world. Wiggling a way through that requires improvisation, but it also requires know-how. Quite a lot of that is relevant to movements. The rest of it is relevant to care.

Protective principles

1. Have an entry strategy. Work out what it is that you want to achieve – in your life and in the world, on whatever timeframe you're comfortable with. Map it out.
2. Look for organisations you feel comfortable in and find ways to make your skills and strengths useful. Collectives are recommended. Check them out online or ask people you know. A good stress-test for the healthiness of an organisation is how it deals with conflict. Ignoring/shutting things down unilaterally can be just as aggressive in practice as shouting someone down. Look for equitable deliberation. Fairness feels right (because it is).
3. Go with someone you know. This makes it easier to step over the barrier into a new space. You can also encourage each other, and help each other with much of the above (although sometimes it can be tempting to fall into constant side-chatting or create a little, separate bubble – try to avoid that, I often forget).
4. Rest. Do not burn out.
5. Keep connected, even if by just making a phone call. Don't isolate yourself. Check in on your friends if they have gone quiet. Invite them to things, even if they have blanked you.
6. Make sure you are having some fun. Movement work doesn't have to be all sombre and sacrificial, no matter how perfectly any 'leaders' are channelling their strictest primary-school teacher. If they are wedded to that caricature, maybe try somewhere else.

7. Avoid power struggles and status bullshit. Talk to people you trust about trying to keep things equitable, horizontal and fair.

8. Commit. Repetition and consistency will mean you can build friendships and forge a sense of belonging. If you can't make something, because of your health or otherwise, try to think of a way you can demonstrate you care even if you are not in the room – for your own sake. If not, be forgiving.

9. Be explicit about your needs and your boundaries. If you have physical or mental-health issues, the group will do their best to accommodate you. It shouldn't always be your responsibility to adapt.

10. Try not to judge others in the group too much. Progressives are pros at creating purist factions and arguing over details that lead to splits. Easier said than done but think about a) whether the process you're in is healthy and constructive, even if you disagree with elements of discussion; and b) whether the overall objective is getting any closer to something actually good. If yes to both, don't have a go at people unnecessarily. If no, jump in.

11. Don't ask yourself if you're doing enough. Ask yourself if you're doing your best (thank you Jennifer).

12. Use your privilege and position to platform others, but don't let it stop you from sharing your perspective. The world you see is different from the world anyone else sees. That said, don't monopolise conversation. Respect the room.

13. Maintain all of the other things that are important to your mental health and wellbeing. If you don't have economic security, it might be difficult to engage as much as you want to in this kind of work. That is not your fault. Some people have solidarity funds to support this stuff, some even get a basic income (like the Hero climate circles). Ask for support in meetings – be it cash, childcare, travel or information.

14. Have an exit strategy. If you are struggling, for whatever reason, it might be best to take a break or change tack.

An exit strategy is a contingency so you can be committed, whilst also having the option to prioritise your care in an emergency. Just as the world will be okay without us, you need to know the group will be okay if you step back.

15. Ask people around you for their advice, both the newbies and the veterans.

We need to dream, awake

Extractivism threatens, disconnects, dominates, poisons, cages and abuses our psyches.

They tell us this is just part of the 'natural' course of human and social evolution. The vast majority of lives are meanwhile disposable, run over by the steam train of what they claim is progress. It's par for the course. Means and ends. Fuel for the furnace. It's human nature, they say, it's immutable.

In 1910, Emma Goldman penned an irate response to this particular brand of bullshit.

Poor human nature, what horrible crimes have been committed in thy name! Every fool, from king to policeman, from the flatheaded person to the visionless dabbled in science, presumes to speak authoritatively of human nature. The greater the mental charlatan, the more definite his insistence on the wickedness and weakness of human nature. Yet, how can anyone speak of it today, with every soul in a prison, with every heart fettered, wounded and maimed? . . .

The experimental study of animals in captivity is absolutely useless. Their character, their habits, their appetites undergo a complete transformation when torn from their soil in field and forest. With human nature caged in a narrow space, whipped daily into submission, how can we speak of its potentialities?

Freedom, expansion, opportunity and, above all, peace and repose, alone can teach us the real dominant factors of human nature and all its wonderful possibilities.

I know Emma was right. More than one hundred years later I can feel this truth encroaching on the edges of our quiet moments. Skimming the surface and dipping its toes in at our most raucous. It really hurts me how little idea we have of who it is we could be. But the promise is sweet.

Many of our mental and physical struggles are embodiments of systems of oppression. It helps to remember that climate change is just one more layer on top of centuries-long struggles. It is an important layer, a potentially life-eviscerating layer. But lives have been eviscerated by this machine for a long time. Those on the sharp end are schooled in how to react strategically. The mad can help. The mad can act.

We need action. But we also need patience. Mark Fisher called it 'revolutionary patience'. I like that a lot. The mad can, perhaps surprisingly to many, be experts in waiting. We are old-timers. This has to percolate through our practices, our visions and our pacing. As adrienne maree brown puts it, we have to 'move at the speed of trust', so as not to harm ourselves or each other.[15] We are already hurt enough.

When, in flashes, we see the exit, we can't just make a break for it. We must make sure everyone is with us, that everyone can make it. Then, moving step by step towards the opening, with the horizon of utopia tantalising, receding beyond the opening, we will experiment with bringing new worlds into being.

Regardless of what everyone else does, regardless of how many people join us on the road, at least we will know how to care for one another in every way that matters. We will fight systems that create sickness and harm. We will imagine alternatives. We will create different futures, learn to tend to the earth and attend to each other's minds. Truly, that's all we need. Thankfully, that's enough.

Call me a madman, but I know another world is possible. We need to dream, awake. Call me a madman if you like. My madness was a long time in the making. Millennia, in fact. Call me a madman. Please, do. It's actually a compliment where I come from. And where we're going to.

Acknowledgements

I WOULD LIKE FIRST AND FOREMOST to thank those without whom I would not be alive today, and without whom I would not know myself. My mum, Careen, has saved my life many times, in many ways, as have my dad, David, and my brothers Joe and George. Without their love, humour, patience, generosity, wisdom and – at times – firmness, it is unlikely that I would have been able to connect with the world again in any intelligible way, let alone experience the aching joy they often bring to my life. I love you all, so much. I also want to thank my extended family, which is too large to list in full, but nonetheless deceptively tight-knit. I especially want to acknowledge, in no particular order: Eden, Margot, Anna, Nichol, Emma, Sarah, Stan, Cathy, Nick, Sue, Richard, Charlotte, Chauncey, Sylvie, Tom, Stef, Claire, Alice, Phoebe, Barbara, Bob and Daniel. There are many others who have helped to support me in important ways throughout my recovery, most notably Chris, Selina, Lia, Oswin, Karen, Julia, Josh, Jessica, Emma, Peter and Siobhan, who acted as loving friends and wise guides. I want to give a special thank you to Jan and Jay, too, for having supported my mum and my brothers when I put them all through a particularly traumatic time.

Despite the numerous slips in my care, largely because of at least twelve years of chronic underfunding, I would be dead were it not for the individuals within and the institution of the UK's

National Health Service. What is being done to the NHS by the Conservatives is criminal, and should be tried as criminal. I know several people who were not as lucky as me and are no longer with us as a result of failures in mental-health coverage, as I know many reading this will, too. I would like to thank, in particular: the ambulance crews who have performed with expert precision and speed; the surgeons who performed life-saving operations at St Mary's hospital, as well as the nurses and other staff who looked after me and put up with me for so long; the physiotherapists at Charing Cross hospital; the prosthetists at the Holderness Limb Fitting Centre; and the ECT practitioners at St Charles. Dr Dietch has been my GP for my entire life and he has consistently gone over and above his level of duty for over a decade – dealing expertly and compassionately with my acute and chronic fears, oscillating crises and at times unfairly challenging demands. Thank you too to all the receptionists, phlebotomists and pharmacists at Lonsdale Medical Centre who have made my care possible over the years. The staff at Bliss Chemist in Kilburn have been an invaluable support, also, providing me with a warm, human space, a real community pharmacy in a city being gutted of these. There is one NHS psychiatrist without whom I might never have left the flat I isolated myself in for two years. She even sometimes paid house visits out of hours to check on me, and to insist I got in the shower and shaved every now and again. Thank you, Dr Ranjit-Singh. I would also like to thank Dr Obuaya and Dr Pfeffer for seeing me in extremis and granting me access to psychiatric hospitals. I promise to pay this forward.

I hold deep gratitude towards my friends, both for putting up with me and for loving me when I was at my best, and at my infuriating and dark worst, as well as letting me talk through the ideas around this book for hours and hours. As sounding boards during the process, and as allies in life, you all make me feel very lucky. I'd especially like to thank James, who first suggested the book as a project, and his partner Lily, as well as my oldest friend Jake, and my friends Jim, Amber, Pete, Max, Nathan, Carla, Cerise,

Ella, Faisal, Imi, Rich, Ira, Will, Tristan, Kweku, Lola, Pato, Alex, Maddy, Dom, Joao, Milo, Alycia, Julia, George, Olya, Ari, Sophie, Rhiannon Karen, Julia, Josh, Jessica, Emma and Martin. These people have added adventure, accountability, elation, comfort, serenity, challenge, intrigue, intelligence, fun and, thankfully, the occasional aggravation.

If it weren't for Leo, I would not have had my eyes opened to the immensity of climate chaos, nor been equipped with the spaces and skills to resist, to reconnect or to remedy. Leo taught me to be the eyes and the ears, to try to see beyond the veil. Over the years I have also been lucky enough to be gifted the support and guidance of mentors whom I can call friends, including through my relationships with George, Ruth, Jamie, Susan and Paul. My endless struggle towards and through post-traumatic growth would not even have become a shimmer on the horizon were it not for the commitment and skill of my recovery coach Jaz and my trauma therapist Ed.

This book was a collaboration, to the extent that I feel strange having my name on the cover (let alone so big). All of the interviewees for this book pushed me into different ways of thinking, but their contributions went deeper than that. Their solidarity and their openness, especially about two issues as raw as climate chaos and mental health, which become even more raw when combined, were radical practices of generosity that I could feel in conversation, in process and in spirit. Working with them all has taught me new ways to be, and new aspects of being I now aspire to. The contributors to Part II especially gave a lot of themselves. Jennifer, Fahad and Emiliano shared not only (a lot of) their time and experience with me, but also trusted me enough to collaborate and struggle together. They took a risk. I hope they feel this book does some justice to their work and to our relationship.

I want to thank all the other interviewees for their faith in me, too, and for their willingness to let me stitch their worlds into what must ultimately be simply my interpretations of their interpretations of reality. Thank you, too, for all the work you do and continue to do. In particular, I appreciate the insights and swerves in perspective I gained from talking to Elaine, Asma, Harpreet, Araceli, Rhiannon, Kelly and Patricio. To all those interviewees who chose to remain anonymous, I respect your allowing such intimate and/or risk-prone content to be put in print, and I promise that I do not underestimate the impact of your stepping over that particular line. To all those of you doing similar work, at whatever geographic scale around the world, I am forever grateful. I hope to see you all on the battlefield.

Many people read and commented on early drafts, principally my dad, David, without whose efforts I would not have pitched this book a second time, let alone completed it, as well as Peter, George and my tireless agent Piers. Many others gave useful critiques and helped me along when I thought this book would never be written, let alone published. Jane and Amanda have been important supports both in terms of understanding my mind and applying some rigour to the amorphous thought blobs occupying my head. So have my rambling chats with George over coffee and cigarettes after chilly morning swims at the Serpentine. The editors at Footnote Press, from David and Kwaku to, later, Candida and Chris, turned an intangible mixture of passionate fury, love and drive into a much more coherent piece of work. Candida's patience with me, as well as her deft skill at making sense out of things I wasn't sure made sense myself, has made the process quick and painless. It has been a joy to work with her and she has given me confidence as a writer that I hope I'm able to hold on to. Thank you to everybody at Footnote for taking a chance on me. I do not take it lightly. (Plus, cheeky little thanks to Gary Lineker for retweeting a post of mine, which gained me enough followers to make me more publishable. Legend.)

Thank you too to *The Ecologist*. Brendan Montague was the first person to publish any of my writing on this topic and he kept printing, guiding and editing new articles as the months went on. That gave me the momentum to continue, the confidence to reach out to people even as I worked out what I felt and thought about the topic and made me feel seen.

Ultimately, my thanks goes to the primordial ooze. The branching diversity of life's incessant drive on this planet is an unending mystery and wonder. I wish I had the ability to sink into that wildness and revel in it, just for a moment. I thank the human and non-human cultures, too, that have kept me alive, inquisitive and adventurous. I thank my body for sustaining me, in the hard times and the good. I thank my mind, too, in both its dynamic equilibrium and in its haunting flare-ups. I also want to thank the wolf. I hope to see you again one day, and in better spirits.

With love and rage, blue skies and starry nights,

Charlie

P.S. I would like to thank, in advance, everyone who will inevitably be affected by any of the negative psycho-social fallout that I will experience following the publication of this book. I am more resilient than I have ever been, but I'll probably end up finding it tough. I struggle with public scrutiny at the best of times, so I (presumptuously) offer my gratitude for your patience and compassion. I will have got things wrong, and I promise this will not have been out of ill-intent or intentional slackness – if you can, please call me in rather than calling me out. I also acknowledge my own responsibility to try to be and stay well, for my sake and yours, but I apologise if I am, at any point, a bit of a dick.

Endnotes

Chapter 1

1. Whilst this phrase is widely used, its origin is disputed. Mark Fisher, for instance, attributes it – or its spirit – to both Frederic Jameson and Slavoj Žižek. Fisher argued that it's easier to imagine the end of capitalism than what comes after it. Very true, but it's worth having a go.
2. grist.org/politics/2019s-biggest-pop-culture-trend-was-climate-anxiety
3. grist.org/language/climate-anxiety-google-search-trends
4. thelancet.com/journals/lanplh/article/PIIS2542-5196(21)00278-3/fulltext#seccestitle130
5. sciencedirect.com/science/article/pii/S0272494422001116?via%3Dihub
6. sciencedirect.com/science/article/abs/pii/S0272494419307145
7. reuters.com/article/climate-change-children-idUSL1N2AV1FF
8. https://tinyurl.com/5yrxn4u6
9. childrenssociety.org.uk/what-we-do/our-work/well-being/mental-health-statistics
10. frontiersin.org/articles/10.3389/fpubh.2022.829674/full#h4
11. https://tinyurl.com/ym4js5up; thelancet.com/pdfs/journals/lancet/PIIS0140-6736(18)31612-X.pdf; who.int/news-room/fact-sheets/detail/depression
12. It was a self-selecting group rather than a controlled study of any sort.
13. Mieko Kawakami, *Breasts and Eggs*, Picador, London, 2021, p. 605

14. https://tinyurl.com/4ex2mvmw; tinyurl.com/bdhru4am
15. ncbi.nlm.nih.gov/pmc/articles/PMC3204264
16. link.springer.com/article/10.1007/s10584-021-03234-6
17. tinyurl.com/2kzdeake; thelancet.com/journals/lanpsy/article/PIIS2215-0366(22)00104-3/fulltext
18. archive.boston.com/lifestyle/green/articles/2009/02/09/climate_change_takes_a_mental_toll/
19. iopscience.iop.org/article/10.1088/1755-1315/1084/1/012007/pdf
20. pbs.org/newshour/nation/climate-change-activists-self-immolation-raises-questions-of-faith-and-protest
21. newrepublic.com/article/164861/need-talk-climate-change-suicide
22. isthishowyoufeel.com/ITHYF5.html#Katrin
23. pubmed.ncbi.nlm.nih.gov/29527590/
24. climatecommunication.yale.edu/visualizations-data/ycom-us/
25. K.M. Norgaard, *Living in Denial: Climate Change, Emotions, and Everyday Life,* MIT Press, Cambridge, UK, 2011
26. pharmaceutical-journal.com/article/news/antidepressant-prescribing-increases-by-35-in-six-years
27. nature.com/articles/d41586-021-02179-1
28. tinyurl.com/3yys2dcr
29. nature.com/articles/d41586-021-00090-3
30. iea.org/articles/global-energy-review-co2-emissions-in-2020
31. tinyurl.com/3yys2dcr
32. iea.org/news/after-steep-drop-in-early-2020-global-carbon-dioxide-emissions-have-rebounded-strongly
33. iea.org/news/global-co2-emissions-rebounded-to-their-highest-level-in-history-in-2021

Chapter 2

1. tinyurl.com/yjb32mce
2. ncbi.nlm.nih.gov/pmc/articles/PMC4119797
3. (which I know is inaccurate, but bear with me)
4. tinyurl.com/2p87m744
5. tinyurl.com/mrvkv9ay
6. The word 'allostatic' comes from the medical term 'allostasis,' defined as 'the process by which a state of internal, physiological equilibrium is maintained by an organism in response to *actual or perceived environmental and psychological stressors* [emphasis added].'

7. karger.com/Article/FullText/510696
8. Gabor Maté, *When the Body Says No,* Vermilion, London, 2019
9. sciencedirect.com/science/article/pii/S2667278221000018?via%3Dihub
10. iberdrola.com/social-commitment/what-is-ecoanxiety
11. atriumclinic.co.uk/wp-content/uploads/2021/08/climate-change-anxiety.pdf
12. Murray Bookchin, *The Ecology of Freedom,* Cheshire Books, Palo Alto, California, USA, 1982
13. theguardian.com/environment/2021/apr/20/climate-emergency-anxiety-threapists
14. tinyurl.com/bdeadmu9
15. h/t Rhiannon Osbourne
16. tinyurl.com/ya9zhcdp
17. DSM-V
18. tinyurl.com/ct9uy36x
19. jahonline.org/article/S1054-139X(21)00568-1/fulltext#sec-sectitle0090
20. As well as disabled people
21. jstor.org/stable/4100723
22. hogg.utexas.edu/the-political-abuse-of-psychiatry-against-dissenting-voices
23. ferris.edu/HTMLS/news/jimcrow/question/2005/november.htm
24. tinyurl.com/4me2mxae
25. ncbi.nlm.nih.gov/pmc/articles/PMC4318286
26. ncbi.nlm.nih.gov/pmc/articles/PMC2800147
27. tinyurl.com/u9r8mhn8
28. Whose most remarkable line is 'I have always valued my lifelessness.'
29. tinyurl.com/2c5jkuy2
30. tinyurl.com/ydh89ju8; tinyurl.com/mskf9kd5

Chapter 3

1. news.uthscsa.edu/study-tracks-unpredictability-of-domestic-violence
2. sciencedirect.com/science/article/abs/pii/S0887618520300517
3. sciencedirect.com/science/article/abs/pii/S0887618520300517
4. who.int/news/item/03-06-2022-why-mental-health-is-a-priority-for-action-on-climate-change
5. Ibid.

6. ncbi.nlm.nih.gov/pmc/articles/PMC9400922
7. apps.who.int/iris/handle/10665/272735
8. tinyurl.com/25f3b7yf
9. Rupa Marya and Raj Patel, *Inflamed*, Penguin, London, 2022
10. dw.com/en/india-why-are-suicides-among-farmers-on-the-increase/a-62991022
11. tinyurl.com/4jzjdau4
12. iopscience.iop.org/article/10.1088/1748-9326/ac22c1
13. climatemigration.org.uk/climate-conflict-syria
14. nature.com/articles/s43247-022-00405-w
15. annualreviews.org/doi/pdf/10.1146/annurev-economics-080614-115430
16. thelancet.com/journals/lancet/article/PIIS0140-6736(19)30934-1/fulltext
17. ncbi.nlm.nih.gov/pmc/articles/PMC3181586
18. savethechildren.org/content/dam/usa/reports/ed-cp/weapons-of-war/weapon-of-war-report-2021.pdf
19. tinyurl.com/y4afnrme
20. pubmed.ncbi.nlm.nih.gov/30866745
21. tinyurl.com/msd7kx9y
22. pubmed.ncbi.nlm.nih.gov/29981991
23. tinyurl.com/25ecs4bj
24. tinyurl.com/4krxemr4
25. ipcc.ch/report/sixth-assessment-report-working-group-ii
26. nytimes.com/2022/08/19/well/mind/heat-mental-health.html
27. ncbi.nlm.nih.gov/pmc/articles/PMC3918032
28. tinyurl.com/4krxemr4
29. newsecuritybeat.org/2015/03/heat-hotheads-effect-rising-temperatures-urban-unrest
30. frontiersin.org/articles/10.3389/fpsyt.2020.00074/full
31. ncbi.nlm.nih.gov/pmc/articles/PMC7310019/
32. onlinelibrary.wiley.com/doi/abs/10.1002/casp.2450040505
33. Ibid.
34. onlinelibrary.wiley.com/doi/abs/10.1002/smi.2615
35. efsgv.org/learn/learn-more-about-gun-violence/mental-illness-and-gun-violence
36. thetrace.org/2021/08/climate-change-gun-violence-shootings-research

37. nejm.org/doi/pdf/10.1056/NEJMsr2028985
38. pubmed.ncbi.nlm.nih.gov/28605987
39. bmcpublichealth.biomedcentral.com/articles/10.1186/s12889-021-12411-2
40. reuters.com/article/uk-southeastasia-haze-idUK-BRE95K0FE20130621
41. asiaone.com/asia/central-kalimantans-psi-pushing-2000
42. bps.org.uk/psychologist/air-pollution-and-mental-health
43. unicef.org/press-releases/indonesia-10-million-children-risk-air-pollution-due-wild-forest-fires
44. Revelations, Chapter 13 in the King James Bible refers to fire falling from the sky.
45. tinyurl.com/4eu77xm7
46. tinyurl.com/bd7vzv75
47. wp.lancs.ac.uk/floodarchive
48. sciencedirect.com/science/article/pii/S0277953620303567
49. pubmed.ncbi.nlm.nih.gov/25974138/
50. Ethan Watters, *Crazy Like Us*, Scribe, Carlton, Australia, 2010, p. 69. site.ebrary.com/id/10644392
51. suicideinfo.ca/local_resource/naturaldisastersandsuicide
52. pubmed.ncbi.nlm.nih.gov/27880626

Chapter 4

1. tinyurl.com/mr85rj6k
2. who.int/news-room/fact-sheets/detail/mental-health-and-forced-displacement
3. pubmed.ncbi.nlm.nih.gov/22982816
4. tinyurl.com/yck6abh6
5. amnesty.org/en/latest/news/2020/06/no-clean-up-no-justice-shell-oil-pollution-in-the-niger-delta/; tinyurl.com/32r46u4r
6. Today there are about 220 million people living in Nigeria. The United Nations Population Fund says it could be home to 400 million by 2050. That's almost double.
7. The event was organised with the help of the Mentally Aware Nigeria Initiative (MANI).
8. tinyurl.com/ypwu34ad
9. oxfam.org/en/nigeria-extreme-inequality-numbers
10. tinyurl.com/2bxhc83u

11. And social-justice spaces of all stripes.
12. tinyurl.com/2p8zrxmw
13. legislation.nsw.gov.au/view/pdf/asmade/act-1958-45
14. tinyurl.com/4bdxtjyf;   thelancet.com/journals/langlo/article/PIIS2214-109X(20)30302-8/fulltext
15. tinyurl.com/ms9t4t5b
16. mhinnovation.net/innovations/aro-primary-care-mental-health-programme
17. hrw.org/news/2019/11/11/nigeria-people-mental-health-conditions-chained-abused
18. tribuneonlineng.com/the-kano-chains; bbc.co.uk/news/world-africa-53893271
19. publichealth.com.ng/list-of-psychiatric-hospitals-in-nigeria
20. bebor.org/wp-content/uploads/2012/09/Ogoni-Bill-of-Rights.pdf
21. websites.umich.edu/~snre492/cases_03-04/Ogoni/Ogoni_case_study.htm
22. tinyurl.com/bdep3ss2
23. tinyurl.com/2t35sn75
24. Amnesty International, *Nigeria: Military Government Clampdown on Opposition*, 11 November 1994, pp 6–7; O. Douglas, 'Ogoni: Four Days of Brutality and Torture', *Liberty*, May–August 1994, p. 22; J. Vidal, 'Born of Oil, Buried in Oil', the *Guardian*, 4 January 1995; W. Soyinka, 'Nigeria's Long Steep, Bloody Slide', *New York Times*, 22 August 1994.
25. justice.gov/eoir/page/file/1014511/download
26. facebook.com/watch/?v=226410348090842
27. issafrica.org/iss-today/endless-oil-spills-blacken-ogonilands-prospects
28. edition.cnn.com/2022/05/25/africa/shell-oil-spills-nigeria-intl-cmd/index.html
29. amnesty.org/en/latest/news/2020/06/no-clean-up-no-justice-shell-oil-pollution-in-the-niger-delta
30. tinyurl.com/yc285ck7 Some compensation is finally rolling in. In 2021 Shell agreed to pay communities $95 million in compensation for oil spills since 1970. The company refused to claim responsibility for the spills, saying they were the work of vandals, thieves and damage done during the Nigerian civil war. The lawsuit was filed way back in 2008 by a group of farmers on behalf of their

community. By the time it reached settlement, two of the initial plaintiffs had died. At the end of 2022, Shell again settled a lawsuit with a group of farmers from the Delta, paying out 15 million euros in response to four spills between 2004 and 2007, at least fifteen years prior. Still, the oil giant insisted that they were making 'no admission of liability'. They were forced to make another similar pay-out in 2015. These figures are utterly dwarfed by the amount Shell spent on defending its assets. Between 2007 and 2009 alone, the company spent $383 million on security, including direct payments to the Nigerian military.

31. thisdaylive.com/index.php/2022/10/17/nema-2-5-million-persons-affected-by-flooding
32. tinyurl.com/2p93u7v9
33. tinyurl.com/3zy2dkvr
34. businessinsider.com/nigeria-major-oil-production-risk-2016-3?r-=MX&IR=T
35. tinyurl.com/7t9cevp5
36. tinyurl.com/yc5wetcx
37. eia.gov/todayinenergy/detail.php?id=27572
38. ijrhss.org/papers/v4-i10/3.pdf
39. aljazeera.com/news/2016/11/13/nigerian-army-presence-prompts-niger-delta-attacks
40. tinyurl.com/3zy2dkvr
41. cfr.org/blog/delegitimizing-armed-agitations-niger-delta
42. tinyurl.com/4xv8y6fm
43. tinyurl.com/327vyvhw
44. This isn't an exaggeration. A recent nationwide poll found that the one symptom most commonly associated with a mental health disorder was 'when someone starts running around naked'. The least connected symptoms were covert, rather than overt, particularly 'when someone starts keeping to themselves', which only a quarter of respondents saw as potentially worrying.
45. shewriteswoman.org/initiatives
46. link.springer.com/article/10.1007/s11126-022-09974-7
47. tinyurl.com/3zmd3pn5
48. https://books.openedition.org/pur/112250?lang=en
49. libpsy.org/wp-content/uploads/2011/11/A-second-psycho-logy-of-liberation.pdf

50. tinyurl.com/5yx3aua2
51. tinyurl.com/yf559dm8

Chapter 5

1. The result is an extremely high-risk situation. Some predict that there could be a 'water war' between the two nations before 2025. Pakistan and India have fought three major wars and two minor wars over Kashmir. In 2019, India instituted the world's longest internet blackout over Kashmir and imprisoned thousands of political opponents following suicide attacks from a Pakistani group. Official Indian government figures say more than 50,000 civilians have been killed in Kashmir since 1989. These disputes are being exacerbated by climate shocks in the region. See, for example: kcl.ac.uk/will-there-be-a-water-war-between-india-and-pakistan-by-2025

2. e360.yale.edu/features/himalayas-glaciers-climate-change; ipcc.ch/report/ar6/wg2

3. This is only made worse by dams like the one recently built by the World Bank which displaced hundreds, as well as badly planned hydroelectric plants, settlements and farms built on riverbanks; together diverting watercourses to devastating effect.

4. tinyurl.com/2p937dxj

5. ourworldindata.org/co2/country/pakistan; dawn.com/news/1520402

6. theguardian.com/world/2022/may/02/pakistan-india-heat-waves-water-electricity-shortages

7. almustafatrust.org/appeals/emergency/pakistan-drought-appeal

8. tinyurl.com/ypuc85wn

9. csis.org/analysis/pakistans-deadly-floods-pose-urgent-questions-preparedness-and-response

10. news.un.org/en/story/2022/09/1126001

11. tinyurl.com/47455vuxl

12. bmj.com/content/328/7443/794

13. tinyurl.com/4yhyscme

14. Ibid.

15. tinyurl.com/t46nkxwm

16. tinyurl.com/2sfkw275; scientificamerican.com/article/psychiatry-needs-to-get-right-with-god

17. al-hakawati.net/Content/uploads/Culture/1293_Islamic_Medicine_Ahead_of_its_Time.pdf

18. tinyurl.com/bdfshdka
19. tinyurl.com/mvwn9uy7; eprints.soas.ac.uk/29435/1/10731591.
    pdf; ncbi.nlm.nih.gov/pmc/articles/PMC8106421
20. ncbi.nlm.nih.gov/pmc/articles/PMC8106421
21. ncbi.nlm.nih.gov/pmc/articles/PMC8977681; tinyurl.com/28xsdj5m
22. caretechfoundation.org.uk/mental-health-in-pakistan
23. Ibid.
24. interventionjournal.com/sites/default/files/Mental_health_and_
    psychosocial_support_for_the.4.pdf
25. tinyurl.com/y6s887sx
26. dawn.com/news/1574424
27. rainforests.mongabay.com/deforestation/2000/Pakistan.htm
28. econ.berkeley.edu/sites/default/files/Ahmed_Ali_thesis.pdf
29. pbs.gov.pk/publication/agricultural-census-2010-pakistan-report;
    tinyurl.com/2p9cduj7
30. ncbi.nlm.nih.gov/pmc/articles/PMC2913558/

Chapter 6

1. radiozapatista.org/?p=12929&lang=en
2. tinyurl.com/2pz7ymp2; Emory Dean Keoke and Kay Marie
   Porterfield, *Encyclopedia of American Indian Contributions to the
   World,* New York, 2002; Bernard Ortiz de Montellano, *Aztec Medicine,
   Health and Nutrition*, Rutgers University Press, New Brunswick, 1990.
3. tinyurl.com/227yvaxh
4. sciencedirect.com/science/article/pii/S0959378020301424
5. gjia.georgetown.edu/2020/05/13/climate-change-in-central-amer-
   ica-the-drug-war-connection/
6. daily.jstor.org/where-drug-trafficking-and-climate-change-collide/
7. scielo.org.mx/pdf/sm/v38n2/v38n2a3.pdf
8. tinyurl.com/2p96fwvb
9. researchgate.net/publication/322371014_Mental_health_services_
   in_Mexico
10. milenio.com/estados/mexico-sequia-84-9-ciento-territorio-nacion-
    al-smn
11. tinyurl.com/y7dhhxuh
12. tinyurl.com/3uatx42w
13. desinformemonos.org/wp-content/uploads/2020/04/Samir-sin-
    reversa.pdf

14. tinyurl.com/yyr3tp94
15. tinyurl.com/ykrtjrwx
16. tinyurl.com/328r3akb
17. nytimes.com/2022/08/17/world/americas/mexico-president-re-newable-energy.html
18. thenation.com/article/archive/mexico-transcanada-pipeline-puebla-indigenous-rights
19. americas.org/yaqui-pipeline-fighters-need-immediate-relief
20. tinyurl.com/32vcm73k
21. bpr.berkeley.edu/2021/08/02/environmental-activism-in-latin-america-comes-with-a-deadly-cost
22. John Berger, *G*, Weidenfeld & Nicolson, London, 1972.
23. radiozapatista.org/?page_id=13233&lang=en
24. Félix Guattari, *The Three Ecologies*, Continuum, London and New York, 2000 (original French edition 1989).
25. Carlos Lenkersdorf, 'The Tojolabal Language and Their Social Sciences', the *Journal of Multicultural Discourses*, Vol. 1, Issue 2, 2006, pp. 97–114, tandfonline.com/doi/abs/10.2167/md015.0
26. eapolanco.com/how-to-pluralize-in-early-modern-nahuatl
27. jstor.org/stable/40316385
28. ajp.psychiatryonline.org/doi/10.1176/ajp.60.2.265
29. psycnet.apa.org/fulltext/2021-94351-001.html
30. rapidtransition.org/stories/the-rights-of-nature-in-bolivia-and-ecuador

Chapter 7

1. medium.com/@rhithink/leaving-schumacher-college-bcda7ee800c1
2. A term coined by her friend Kathleen Cassidy to undermine and usurp the patriarchal overtones of conventional graduate academia.
3. nature.com/articles/s41558-018-0102-4
4. medium.com/@rhithink/leaving-schumacher-college-bcda7ee800c1
5. Sabine Hossenfelder, *Existential Physics*, Penguin Random House, 2022
6. tinyurl.com/4263mfk5
7. jofreeman.com/joreen/tyranny.htm
8. Erik Olin Wright's styles of action for systemic change, for instance, are: i) interstitial; ii) ruptural; and iii) symbiotic transformation. The Three Horizons methodology uses, well, three horizons . . . etc.
9. fao.org/news/story/en/item/197623/icode

Chapter 8

1. guerrillagardening.org
2. seedballskenya.com/throw-grow
3. spectator.co.uk/article/the-scourge-of-the-grouse-moor
4. https://tinyurl.com/2p93ymff
5. insider.com/tiktok-seed-bombing-trend-is-radical-and-illicit-creators-say-2021-7
6. globalclimatepledge.com/zip-code-matters-racial-and-economic-disparities-in-green-space
7. tinyurl.com/234knu54
8. tinyurl.com/sybjztjd
9. rollingstone.co.uk/politics/features/tyre-extinguishers-suv-4x4-climate-action-campaign-21323
10. Andreas Malm, *How to Blow Up a Pipeline,* Verso, London, 2021.
11. tyreextinguishers.com/how-to-deflate-an-suv-tyre
12. tyreextinguishers.com/leaflet
13. tinyurl.com/tcarxn3n
14. mylondon.news/news/south-london-news/im-62-year-old-mum-25176768
15. tinyurl.com/4zuhcaf3
16. weforum.org/agenda/2022/10/fossil-fuels-incompatible-1-5c-goal-energy-climate-change-study
17. reuters.com/business/environment/un-chief-says-dash-new-fossil-fuels-is-delusional-2022-06-14
18. Podcast with Andreas Malm.
19. bbc.com/news/uk-england-60951403
20. world.350.org/pacificwarriors/2014/10/20/coal-ships-stopped-the-warriors-have-risen
21. aclu.org/issues/free-speech/rights-protesters/stand-standing-rock
22. tinyurl.com/57ttu7tk
23. indigenousclimateaction.com/entries/new-report-indigenous-resistance-against-carbon
24. nplusonemag.com/issue-43/essays/migizi-will-fly
25. Ibid.
26. tinyurl.com/4n3ckn52
27. tinyurl.com/2tm2mfxp
28. Derrick Jensen, *Endgame. Volume 2: Resistance,* Seven Stories Press, New York, 2006, p. 813.

29. climateculture.earth/directory/bad-activist-collective
30. https://linktr.ee/SeanDaBlack
31. archive.org/details/hackthiszine1/page/n3/mode/2up
32. https://tinyurl.com/yc7xsvbd
33. As her Twitter bio puts it.
34. In total, 152 Just Stop Oil protesters have been jailed since the group's inception. See: tinyurl.com/3nru8mkz
35. Fittingly for an artistic movement, the term 'art-tivism' is good to look at but not as easy to have roll off the tongue.
36. brandalism.ch/wp-content/uploads/2016/12/Brandalism_Subvertising_Manual_web.pdf
37. weareadg.org This is not a paid (or unpaid) promotional endorsement – honestly, they're just wicked. I'd also recommend MOCCAM – tinyurl.com/2p8fdhtk

Chapter 9

1. ncbi.nlm.nih.gov/pmc/articles/PMC6169872; frontiersin.org/articles/10.3389/fpsyg.2022.825161/full
2. commonslibrary.org/deep-canvassing-scripts-and-examples
3. tinyurl.com/99366chn
4. ncbi.nlm.nih.gov/pmc/articles/PMC4302252
5. tinyurl.com/5n6zs593
6. tinyurl.com/afwwvhe8
7. yali.state.gov/courses/course-942/#
8. commonslibrary.org/organising-in-a-pandemic
9. All the rules available here, courtesy of Open Culture: openculture.com/2017/02/13-rules-for-radicals.html
10. wcl.nwf.org/wp-content/uploads/2018/09/Marshall-Ganz-People-Power-and-Change.pdf
11. sdinet.org/who-is-sdi/about-us/
12. tinyurl.com/3yms5jh2
13. tinyurl.com/3hd5bycb; facebook.com/mahilamilan
14. toamazonia.org/cases-studies
15. cer.org.za/programmes/activist-support-training/training
16. Inspired by Bernie Sanders' 'barnstorming' techniques, we tried to organise a national 'bargestorming' operation, taking advantage of the fact that people living on houseboats can vote in any constituency – so could powerfully influence swing seats. The

campaign didn't materialise but I still think it's a great idea for the future.

17. Particularly through Momentum, The World Transformed, unions and satellite organisations.
18. instructionalcoaching.com/paulo-freires-five-ideas-for-dialogical-learning
19. littlesis.org
20. beautifultrouble.org/toolbox/tool/power-mapping
21. littlesis.org/oligrapher/1634-who-s-banking-on-the-dakota-access-pipeline
22. tinyurl.com/39sssnnx
23. talkofftherecord.org/media/1090/mind_the_gap_web.pdf
24. blackmindsmatteruk.com
25. theblackmensconsortium.com/podcasts
26. tinyurl.com/pwhwp5xd
27. kokocares.org
28. ctb.ku.edu/en/table-of-contents/implement/enhancing-support/peer-support-groups/main
29. commonslibrary.org/deep-canvassing-to-shift-hearts-mind-and-votes/#Script_Development
30. As the Bernie Sanders campaign successfully managed. See: Chapter 7 of Becky Bond and Zack Exley, *Rules for Revolutionaries*, Chelsea Green, White River Junction, VT, 2017.
31. foodfoundation.org.uk/press-release/alarming-increase-food-in-security-now-affecting-four-million-children
32. sustainweb.org/news/jan18_calls_grow_for_government_food_insecurity_measurement
33. nationalfoodservice.uk/branches
34. theguardian.com/commentisfree/2017/feb/08/take-back-control-bottom-up-communities
35. goodlawproject.org/update/rare-and-vital-win
36. goodlawproject.org/case/net-zero-the-government-must-take-the-climate-crisis-seriously
37. clientearth.org/our-global-reach/africa/gabon
38. When Feliciano died, he refused to be cremated because he thought it was a waste. He believed he could produce a good crop of apples, his partner Martin told me, but the law stipulated that coffins had to be encased in cement. There is an apple tree on Feliciano's grave but its roots have hit cement.

39. ourchildrenstrust.org/juliana-v-us
40. climatecasechart.com/non-us-case/future-generation-v-ministry-environment-others
41. tinyurl.com/ykjb5bam
42. newspapers.com/clip/47593344/the-indianapolis-star
43. tinyurl.com/bdzz8py8
44. A joke, an oldie but goodie: Q: How many therapists does it take to change a lightbulb? A: One, but the lightbulb has to want to change itself.
45. tinyurl.com/499xa5b4
46. journals.sagepub.com/doi/10.1177/0899764018783277
47. greattransition.org/publication/liberation-ecology
48. tinyurl.com/yhrzuras
49. crcc.usc.edu/fazlun-khalid-environmentalism-is-intrinsic-to-islam
50. tinyurl.com/4u27t6ms
51. teapartypatriots.org/attend-school-board
52. pure.coventry.ac.uk/ws/portalfiles/portal/29909807/Binder2.pdf
53. tinyurl.com/45j7nn5k
54. tinyurl.com/mrx9rfun
55. A term I like for this, introduced to me by my friend Matt Megary, is 'meso-politics.'
56. righttoroam.org.uk; mobile.twitter.com/everyonesstars
57. tinyurl.com/5bavnwwy
58. twitter.com/CIVIC_SQUARE
59. thefrontroom.cc/about
60. cyclinguk.org/article/rose-lamartine-yates-cycling-uk-suffragette
61. jstor.org/stable/27648112; nabss.org.uk
62. leftbookclub.com
63. abolitionistfutures.com/reading-groups; they also do mutual-aid training.

Chapter 10

1. tinyurl.com/bp5vp8cv
2. onlinelibrary.wiley.com/doi/10.1111/1758-5899.12827
3. magazine.scienceforthepeople.org/geoengineering/land-use-geoengineering-disposession
4. tinyurl.com/4mzphmnp
5. washingtonpost.com/climate-environment/2023/01/09/make-sun-sets-solar-geoengineering-climate

6. My own calculation, using data from the following sources: tinyurl.com/5y9c6d6w; tinyurl.com/5n7wjyzv; producer.com/crops/improving-the-internal-combustion-engine

7. tinyurl.com/3auf7ckt

8. tupa.gtk.fi/raportti/arkisto/42_2021.pdf

9. It is already causing conflict and environmental devastation in South America.

10. tinyurl.com/yckde588

11. tinyurl.com/bdf7vhh9

12. tradingeconomics.com/commodity/lithium

13. nytimes.com/2015/06/28/magazine/i-dont-believe-in-god-but-i-believe-in-lithium.html

14. I worked out once that it would take me 120 years of taking it at my dosage to consume as much lithium as is in one EV battery.

15. Also referred to as a 'Green Industrial Revolution' and similar names across the world.

16. youtube.com/watch?v=d9uTH0iprVQ

17. See: Steady State Economics and Degrowth, blossoming fields in economics dedicated to figuring out how to step off the growth treadmill.

18. In the Erik Olin Wright 'Envisioning Real Utopias' sense.

19. From Freud's essay 'A Metapsychological Supplement to the Theory of Dreams' (1917).

20. standard.co.uk/news/london/boris-johnson-living-with-covid-rules-axed-b983726.html

21. tinyurl.com/3mpzv5ku

22. Pyotr Alexeyevich Kropotkin, *Mutual Aid : A Factor of Evolution*, Penguin, London, 2022 [1902].

23. deanspade.net/2019/12/04/mutual-aid-chart

24. regenerationmag.org/mutual-aid-a-factor-of-liberalism

25. mutualaid.wiki

26. involve.org.uk/resources/blog/opinion/citizens-assembly-behind-irish-abortion-referendum

27. vox.com/2015/6/9/8751267/iceland-capital-controls

28. climateassembly.uk/recommendations/index.html

29. appropedia.org/List_of_climate_assemblies

30. tinyurl.com/yc7jbtk6; globalassembly.org

31. loomio.com/g/BiOwvK4d/xr-people-s-assemblies
32. knoca.eu
33. jstor.org/stable/23047817
34. rapidtransition.org/stories/the-rights-of-nature-in-bolivia-and-ecuador/
35. theguardian.com/world/2022/sep/05/chile-votes-overwhelmingly-to-reject-new-progressive-constitution
36. participedia.net/case/4334
37. thersa.org/blog/2022/10/can-a-universal-basic-income-support-mental-health
38. bostonglobe.com/2021/06/01/opinion/biden-should-impose-carbon-fee-immediately
39. tinyurl.com/57rkcrtk
40. energyinnovationact.org/how-it-works
41. nature.com/articles/s41558-021-01217-0
42. tinyurl.com/2vyhn92e
43. nytimes.com/2022/09/15/us/california-cities-guaranteed-income.html
44. givedirectly.org/ubi-study
45. herocircle.app
46. tinyurl.com/j96jnskv
47. M. Virtanen, S.A. Stansfeld, R. Fuhrer, J.E. Ferrie and M. Kivimäki, 'Overtime Work as a Predictor of Major Depressive Episode', *PLOS ONE*, 7(1), 1–5, 2012 ; link.springer.com/article/10.1007/s10902-018-0008-x ; asahi.com/ajw/articles/14469361
48. tinyurl.com/42xndy27
49. linktr.ee/thenapministry
50. reset-uk.org/static/HowToResetReport-cee340c5dc0708198990294843c4ee28.pdf
51. ips-journal.eu/regions/europe/why-women-deserve-a-four-day-week-4610
52. tinyurl.com/s8m3n366
53. 4dayweek.co.uk ; 4dayweek.co.uk ; autonomy.work/ourwork
54. action.4dayweek.com/for-employees
55. fourdayweek.co.uk ; 4dayweek.io/companies
56. tinyurl.com/yc6yp6vh
57. earth.stanford.edu/news/science-behind-extinction
58. ourworldindata.org/life-on-earth ; tinyurl.com/p94zup8d
59. explorer.land/p/organization/gra/organizations

60. tinyurl.com/4h92syba
61. rewildingbritain.org.uk/support-rewilding/our-campaigns-and-is-sues/climate-emergency
62. rewildingeurope.com/news/rewilding-europe-launches-ambitious-new-strategy-for-2030
63. tinyurl.com/bdm2fc9r
64. https://www.rewildingbritain.org.uk/start-rewilding/12-steps-to-rewilding
65. Giving rights to nature and our non-human kin is a core part of reconnecting with equity and repositioning our species in relation to the rest of life on earth. In his book *Half-Earth* (Liveright Books, New York, 2016), legendary biologist E.O. Wilson put forward a radical yet pragmatic proposal to combat plummeting biodiversity: give half of the land exclusively over to other species. I find the idea exhilarating.
66. investmentpolicy.unctad.org/international-investment-agreements/treaty-files/2775/download
67. tandfonline.com/doi/abs/10.1080/00358539408454216?journal-Code=ctrt20
68. press.un.org/en/2022/ga12482.doc.htm
69. sciencedirect.com/science/article/pii/S2542519620301960
70. doughnuteconomics.org/tools/11
71. Join today at: doughnuteconomics.org/discover-the-community
72. tinyurl.com/2y6cuetx
73. twitter.com/ImmyKaur/status/1602739647448219648
74. P. Cullors on podcast: Abolition is For Everybody, Abolition is For Everybody, S1 E2, 2021.
75. Available for free online here: global-gnd.com

Chapter 11

1. The very fact that I distinguish between the two – 'humans' and 'the earth' – is evidence of a cleaving apart, one that survives and continues replicating itself through language. The earth, after all, includes us.
2. Murray Bookchin, *The Ecology of Freedom,* Cheshire Books, Palo Alto, California, USA, 1982
3. David Graeber and David Wengrow, *The Dawn of Everything*, Allen Lane, London, 2021.

4. stopcambo.org.uk/updates/climate-public-health
5. This is deeply democratic, in the truest sense of the word. In their brilliant book *The Dawn of Everything* (see note 3 above), David Graeber and David Wengrow outline that the origins of egalitarianism, participatory democracy and other forms of engaged social organisation lie not in the Enlightenment and the works of people like Jean-Jacques Rousseau but in the critical reflections of indigenous people encountering Western civilisation, like the Wendat philosopher Kandiaronk, whom the more famous scholars simply emulated.
6. healingjusticeldn.org/resources/the-power-of-somatics-for-collective-transformation
7. libcom.org/article/mental-patients-union-1973
8. sanitybytanmoy.com/the-fish-pamphlet-turns-50
9. twitter.com/climateincolour/status/1617961602330030082?s=20&t=4PQF8i2ZwDSJJtsZVUWtVQ
10. tinyurl.com/yc8yzw55
11. blacklivesmatter.com/healing-justice-webinar
12. Somatics has roots in many other cultures, from Himalayan Buddhism and Ayurveda to Celtic and Viking traditions and the varied shamanisms of the Australasian, American and African continents.
13. Staci K. Haines, *The Politics of Trauma*, North Atlantic Books, Berkeley, USA, 2019.
14. Crip skills, as a framework, is inspired by Kim Katrin Milan's model of Femme Science. For more information see Chapter 2 of: Leah Lakshmi Piepzna-Samarasinha, *Care Work: Dreaming Disability Justice,* Arsenal Pulp Press, Vancouver, Canada, 2021.
15. yesmagazine.org/social-justice/2018/03/27/the-world-is-a-miraculous-mess-and-its-going-to-be-alright

Index

Good jumping-off points ...

All We Can Save, **Force of Nature** and **Tipping Point** are each brilliant climate and mental-health organisations, with lots of resources for learning about and joining communities of action and care, as is the **Climate Journal Project** (which is largely for young people); **The Resilience Project** is a similar, UK-based, project that also provides internationally applicable guidance for practical help (see the *Resources* tab on their website); **The Climate Psychology Alliance** and **The Climate Psychiatry Alliance** offer online peer-support spaces and therapies with climate-aware practitioners (as well as directories of therapists); **The Good Grief Network** has a 10-step support program, available online from anywhere in the world (although it's not free); **Project Inside Out** offers a fresh perspective on all of this, encouraging us to process despair by becoming 'guides'. More academic but still action-based networks include **COP Squared**, **Land Body Ecologies** and **Climate Cares**; Britt Wray's **Generation Dread** newsletter has become a brilliant and lively community for the likeminded climate and mental health aware. **Gen Dread** and **All We Can Save** also pulled together a list of tips and resources for addressing climate change emotions (see: *allwecansave.earth/emotions*), similar, but distinct, from the 'eco-emotions' guide offered by **Climate Cares**. (There is more on explicit mental-health support towards the end of this section.)

Organising networks, strategies and toolkits are everywhere. First and foremost, please put this down at the end of this sentence and check out the utterly astonishing material from **Beautiful Trouble** and **Beautiful Rising**. The collective skill, experience and love involved, let alone their prodigious output, deserves everyone's attention. The site, and the books/cards/trainings etc, will point you to organisations all over the world, such as **The Resistance Hotline**, where activists are invited to call an 0800 number (US), or post questions online for help with nonviolent direct-action tactics, advice and trainings.

There's also the online repository **Community Tool Box**, a wonderful **Organising School** from **Tipping Point** and **Changemakers**, as well as online courses from the **Workers' Educational Association**. Lots of the ideas covered (and not covered) in this book's Chapter 10 'Remedy' can be researched further using online platforms like **Participedia, Evonomics, The Alternative,** and through organisations like the **New Economics Foundation** and **The Democracy Collaborative**, especially the latter's **Next System Project**, plus **Centric Lab** and **Dark Matter Labs** if you like systems work. The likes of **Little Sis**, the **Autonomous Design Group**, **The CreaTures CoLaboratory** and Hacktivist collectives like **Guacamaya** (who published two tera-bytes of mining-company emails in 2022) are also great sources of practical ideas and possible collaborations.

To learn by doing, together, **Healing Justice London** hold free, open events online as well as in-person in London, UK, as does the Nigeria-based **The Eco-anxiety in Africa Project**. **Civic Square** in Birmingham, UK, is place-based, providing physical space for community visioning and action, plus shareable open-use tools available on their site. In addition to those mentioned in the book, some umbrella groups worth flagging include **Climate Action Network, Indigenous Climate Action,** the **Global Campaign to Demand Climate Justice, Earth Guardians** and **350.org**. All can help you find local groups, and pointers to others. Generally, community-organising outfits, climate justice and mental health-aware direct-action groups and mutual-aid networks are good to search for (i.e. outward facing, healthy collective action).

For individual and collective mental-health support, the charity **Mind** (UK) provides supported spaces, online and IRL (see their *Peer Support* page), as do outfits like the **Hearing Voices Network**, the free worldwide app **Koko,** and spaces like **#Psychosischat** on Twitter. If you don't know where to look locally, odd as it may sound, it's worth approaching healthy-looking mental-health communities and organisations on Twitter, Reddit, Instagram etc; **Mentally Aware Nigeria Initiative** and **She Writes Woman** provide Africa-wide support. Many of these organisations have lots of useable material on their websites, plus links to other organisations and therapies covering different parts of the world. I have personally found **recovery coaching** (solution-focused pragmatic, structured help that's lighter on the hyper-analytical chat) as well as **trauma therapy** and **somatic therapy**, immensely helpful. Everyone's needs are different.

This author was helped by learning about liberation psychology, systemic therapies and (cautiously) the whole universe of anti-psychiatry and associated, but often distinct, campaign work of psychiatric patients and disabled allies. The books *The Politics of Trauma*, *Care Work: Dreaming Disability Justice*, and *Pleasure Activism,* as well as *The Body Keeps the Score, Inflamed: Deep Medicine and the Anatomy of Justice,* and *Rest is Resistance* were of great help. They provided me with safety tips, visioning and drive. For autobiographical insights from inside madness, I also value and recommend (if you're feeling resilient enough): Jay Griffiths' *Tristimania*, Esmé Weijun Wang's *The Collected Schizophrenias* and Zoe Thorogood's graphic novel *It's Lonely at the Centre of the Earth.*

If utopias are just for 'walking', they nonetheless have immense utility. Real utopian ideas make up an ecosystem of different, inter-related visions that draw us towards systemically better futures. Whether or not they agree on key features is less important than the process of diverse visions contributing to multifaceted experimentation. We learn as we go. These can also help secure our confidence that we are not mad (in the derogatory colloquial sense), but justified in our anger, disquiet and despair. I have been helped by the illuminating ambition and clarity of books like *Envisioning Real Utopias, Becoming Kin, Angela Y. Davis: An Autobiography, To Struggle is Human, Mutual Aid* (Dean Spade's and Kropotkin's), *Capitalist Realism, Pedagogy of the Oppressed, Black Skin White Masks, Planet on Fire, Half-Earth Socialism, Planetary Politics, Degrowth and Strategy, Endgame, Beyond State Power and Violence, Rules for Revolutionaries, The Three Ecologies, PARECON, The Ecology of Freedom, The Divide, Fully Automated Luxury Communism* and *Wobblies and Zapatistas*. The podcasts *Movement Memos, UpStream, For The Wild, Mother Country Radicals, Climate Crisis Conversations, Green New Deal Media, ACFM,* and *Liberty Tactics* are also great sources of inspiration, support, comfort and further connection. My fellow activists Tori Tsui and Mikaela Loach both have incandescent new books out: *It's Not Just You* and *It's Not That Radical*, respectively. Finally, I gained a lot of much-needed and relevant reflective space from Leo Tolstoy's *Resurrection*, and NK Jemisin's *Broken Earth* series, which I read in hospital. The novels explore themes of moral, economic, ecological, spiritual and emotional justice, from the personal to the universal, and all the inevitable fuck-ups that ensue. We're only human. See you out there . . .

About the author

Charlie Hertzog Young is a researcher, writer and award-winning activist. A proudly mad bipolar double amputee, he has worked for the New Economics Foundation, the Royal Society of Arts, the Good Law Project, the Four Day Week Campaign and the Centre for Progressive Change, as well as the UK Labour Party under three consecutive leaders. Charlie has spoken at the LSE, the UN and the World Economic Forum. He has been an activist and organiser since his teens, has co-founded two organisations and was involved in staging a widely publicised walk-out of Gregory Mankiw's economics lectures, the Harvard professor who was George W. Bush's chief White House economic advisor. As well as Harvard, he studied at the School of Oriental and African Studies and at Schumacher College, and has written for *The Ecologist*, the *Independent*, *Novara Media*, *Open Democracy* and the *Guardian*.